RaumFragen: Stadt – Region – Landschaft

Series Editors

Olaf Kühne, Forschungsbereich Geographie, Eberhard Karls Universität Tübingen, Tübingen, Germany

Sebastian Kinder, Forschungsbereich Geographie, Eberhard Karls Universität Tübingen, Tübingen, Germany

Olaf Schnur, Stadt- und Quartiersforschung, Berlin, Germany

RaumFragen: Stadt – Region – Landschaft | SpaceAffairs: City – Region – Landscape

Im Zuge des „spatial turns" der Sozial- und Geisteswissenschaften hat sich die Zahl der wissenschaftlichen Forschungen in diesem Bereich deutlich erhöht. Mit der Reihe „RaumFragen: Stadt – Region – Landschaft" wird Wissenschaftlerinnen und Wissenschaftlern ein Forum angeboten, innovative Ansätze der Anthropogeographie und sozialwissenschaftlichen Raumforschung zu präsentieren. Die Reihe orientiert sich an grundsätzlichen Fragen des gesellschaftlichen Raumverständnisses. Dabei ist es das Ziel, unterschiedliche Theorieansätze der anthropogeographischen und sozialwissenschaftlichen Stadt- und Regionalforschung zu integrieren. Räumliche Bezüge sollen dabei insbesondere auf mikro- und mesoskaliger Ebene liegen. Die Reihe umfasst theoretische sowie theoriegeleitete empirische Arbeiten. Dazu gehören Monographien und Sammelbände, aber auch Einführungen in Teilaspekte der stadt- und regionalbezogenen geographischen und sozialwissenschaftlichen Forschung. Ergänzend werden auch Tagungsbände und Qualifikationsarbeiten (Dissertationen, Habilitationsschriften) publiziert.

Herausgegeben von
Prof. Dr. Dr. Olaf Kühne, Universität Tübingen
Prof. Dr. Sebastian Kinder, Universität Tübingen
PD Dr. Olaf Schnur, Berlin

In the course of the "spatial turn" of the social sciences and humanities, the number of scientific researches in this field has increased significantly. With the series "RaumFragen: Stadt – Region – Landschaft" scientists are offered a forum to present innovative approaches in anthropogeography and social space research. The series focuses on fundamental questions of the social understanding of space. The aim is to integrate different theoretical approaches of anthropogeographical and social-scientific urban and regional research. Spatial references should be on a micro- and mesoscale level in particular. The series comprises theoretical and theory-based empirical work. These include monographs and anthologies, but also introductions to some aspects of urban and regional geographical and social science research. In addition, conference proceedings and qualification papers (dissertations, postdoctoral theses) are also published.

Edited by
Prof. Dr. Dr. Olaf Kühne, Universität Tübingen
Prof. Dr. Sebastian Kinder, Universität Tübingen
PD Dr. Olaf Schnur, Berlin

More information about this series at http://www.springer.com/series/10584

Karsten Berr · Olaf Kühne

"Und das ungeheure Bild der Landschaft…"

The Genesis of Landscape Understanding in the German-speaking Regions

 Springer VS

Karsten Berr
Geographisches Institut
Eberhard Karls Universität Tübingen
Tübingen, Germany

Olaf Kühne
Geographisches Institut
Eberhard Karls Universität Tübingen
Tübingen, Germany

ISSN 2625-6991   ISSN 2625-7009   (electronic)
RaumFragen: Stadt – Region – Landschaft
ISBN 978-3-658-27755-0   ISBN 978-3-658-27756-7   (eBook)
https://doi.org/10.1007/978-3-658-27756-7

This Springer VS imprint is published by the registered company Springer Fachmedien Wiesbaden GmbH part of Springer Nature.
The registered company address is: Abraham-Lincoln-Str. 46, 65189 Wiesbaden, Germany

Contents

1   Introductory Remarks   1

2   Theoretical Framing: The Creation of Landscape   3
    2.1   Fundamental Considerations on the Societal and Individual
          Construction of Landscape and Its Relationships
          to Substantive Foundations   3
    2.2   The Importance of Words and Language in the Creation
          of the World   8

3   The Genesis of the Landscape Concept in the German Language
    Regions – The Common Sense Understanding.   11
    3.1   On the Classical Prehistory of the Concept of Landscape –
          Origins Beyond the German-Speaking World   12
    3.2   Medieval Roots in the German-Speaking World   13
    3.3   The European Influence I: Aestheticisation in the Renaissance   16
    3.4   The European Influence II: Between Idealisation and Reification
          of the Landscape – The Landscape Garden   19
    3.5   A German Affair I: The Influence of Romanticism   22
    3.6   A German Affair II: Landscape and 'Heimat' –
          The 'Heimatschutzbewegung' (English Approximately:
          Homeland Security Movement)   27
    3.7   The Understanding of Landscape Today   31
    3.8   Bottom Line   36

4   The Concept of Landscape in Landscape-Related Sciences   37
    4.1   Introductory Remarks   37
    4.2   Classics on the Discovery of the 'Landscape' and a Classic
          German Scientist Dispute   38
    4.3   Classics on the Conceptual History and Historicity of 'Landscape'   41
    4.4   Recent Discussions   46

4.5   A German Special Path: Landscape in German Geography,
      Illustrated by Comparison with Anglo-Saxon Geography   49
4.6   Bottom Line   54

5   **Landscape Research in Its Interdisciplinary and Transdisciplinary**
    Challenge   57
    5.1   Introductory Remarks   57
    5.2   Disciplinary and Interdisciplinary Landscape Research   58
    5.3   Interdisciplinary Landscape Research   59
    5.4   Transdisciplinary Landscape Research   63
    5.5   Conclusion and Outlook   65

6   **Bottom Line**   67

References   71

Introductory Remarks

"Und das ungeheure Bild der Landschaft…" (English approximately: "And the immense image of the landscape…") the second verse of the poem *Mächtige Landschaft* (English: "mighty landscape") by the German poet Christian Morgenstern (1871–1914) summarises in six words many aspects of the complex understanding of landscape in German: Their cultural effectiveness (countless poems have been written about landscape), their oscillation between image and object, their potential to impress, to relativise their own meaning. In this book, we would like to introduce the international public to essential aspects of the specific shaping of the relationship between society and landscape as it has developed in the German-speaking world since the Middle Ages.

Landscape is created constitutively by three components: the individual, the societal, and the physical. Without individual persons, social agreements of landscape could not be updated, without the society, the individual could not form an idea of landscape, without the physical the individual and the society lacked the stimulus to composite together objects into landscape. Since all three components are not immutable, but are subject to changes, they can be described as a 'threefold landscape transformation' (Kühne 2019c). Theoretical approaches to dealing with landscape ultimately integrate these three components in different ways and with different weighting, at times ignoring one or another. These three components are integrated not only by scientific theories, but also by non-scientific ideas of landscape, whereby there can be great cultural differences between the weighting of the individual components, but also between their connections (among many: Bruns et al. 2015; Bruns and Kühne 2013; Drexler 2010; Olwig 2002).

The concept of landscape in the German-speaking world has been addressed in English literature (e.g. in Cosgrove 2004; Olwig 1996, 2002; Antrop 2018; Kühne 2015a, 2018d), yet a detailed and *common sense understanding* as well as scientific understanding of landscape including representation is missing so far. In this respect, this gap is now to be closed by preparing and evaluating the extensive German-language literature on this topic (to be mentioned by way of example: Berr 2019b; Berr et al. 2019;

K. Berr and O. Kühne, *"Und das ungeheure Bild der Landschaft…"*, RaumFragen: Stadt – Region – Landschaft, https://doi.org/10.1007/978-3-658-27756-7_1

Berr and Schenk 2019; Drexler 2009b; Eisel 1982; Hard 1969, 1970a, c, 1973, 1977; Hartlieb von Wallthor and Quirin 1977; Körner 2005a, b, 2017; Kühne 2018e, f; Müller 1977). The differentiated examination of the German-language concept of landscape does not only appear to be of interest to the international public in terms of conceptual history, it also brings forth a contribution to clarifying the contingency of landscape concepts (and more generally: abstract concepts). The examination of German-language landscape concepts also benefits the international public, since German-language landscape research from the turn of the nineteenth century to the twentieth century provided essential impulses for the terminological version of 'landscape' (see e.g. Mathewson 2009). In this respect, the present book is aimed in particular at people in the various landscape-related sciences (such as geography, landscape architecture, landscape planning, but also architecture, archaeology, sociology, anthropology, philosophy, etc.) who deal with aspects of origin, change of meaning, and exchange of meaning associated with the term 'landscape'.

In accordance with the thematic version of this book, we pursue a constructivist basic perspective (among other things in connection with the following: Cosgrove 1984; Cosgrove and Jackson 1987; Greider and Garkovich 1994; Kühne 2018f, 2019f; detailed in Chap. 2). With regard to 'threefold landscape transformation', the focus of the work is not on the material foundations of landscape, but on the components of social construction as well as the individual influence upon landscape understanding (especially in and through science).

This book deals, firstly, with the specific development of the German-language concept of landscape as an achievement of integration of these three components in the context of the development of a common sense understanding of landscape (Chap. 3). It deals, secondly, with the development of the scientific concept of landscape in German-language scientific discourses (which certainly differ from one subject to another; Chap. 4). Thirdly, however, it also deals with the interrelations between scientific and non-scientific conceptions (Chap. 5). Fourthly, a comparison is also endeavored between German-language landscape perceptions (scientific and non-scientific) and those with different cultural ties (Sects. 3.7 and 4.5). In doing so, it is essentially a matter of clarifying the German-language discussions and specifics, whereby the overall logics of the creation of landscape concepts are to be broached. However, before these objects can be presented, the theoretical framework of the examination is dealt with, firstly to define the terminology used, secondly to prepare key processes in the creation of landscape, thirdly to clarify the scientific perspectives of the authors (Chap. 2).

Theoretical Framing: The Creation of Landscape

In this chapter, the analytical fundamentals for the comprehensions of landscape in the German-speaking regions illustrated in the following chapters will be presented. In addition, the theoretical framing of the investigation (constructivism) is presented and its rational explained. Since language is of great significance in the social fabric of the world, its connection to the theme of 'landscape' will be discussed in more detail.

2.1 Fundamental Considerations on the Societal and Individual Construction of Landscape and Its Relationships to Substantive Foundations

The scientific grasp of the world in general, and towards landscape in particular, take place along the three-dimensional intersect of the individual, societal, and substantive mentioned in the introduction. Ontological, ethical, and aesthetic modes of observation are specifically linked with epistemological questions and (at least partially) provided with normative statements. In the history of scientific landscape research, four paradigmatic underlying positions can be identified, which will be referred to in the following – as an example, in the history of landscape research in Germany or in comparison to other lines of research tradition. These are essentialist, positivist, constructivist, and intermediate positions (more detailed e.g. in Kühne 2018f, 2019c; Winchester et al. 2003; Wylie 2007; Duncan and Duncan 2009).

The basic assumption of an essentialist understanding of landscape is the assumption that it has its own 'essence'. The external, material manifestations are determined by this nature, how regional (=landscape-specific) cultural specifics constitute the 'essence' of this landscape. Landscape and culture mutually imprinted upon each other. In addition to these 'essential' properties, i.e. properties that would correspond to the 'essence' of landscape-culture symbiosis, there are 'accidental' elements (material objects in relation

© Springer Fachmedien Wiesbaden GmbH, ein Teil von Springer Nature 2020
K. Berr and O. Kühne, *"Und das ungeheure Bild der Landschaft…"*, RaumFragen: Stadt – Region – Landschaft, https://doi.org/10.1007/978-3-658-27756-7_2

to landscape or modes of action, interpretations, and evaluations in the context of the 'regional community') that would occur accidentally there, would not correspond to the 'essence' (more detailed: Albert 2005; Chilla et al. 2015, 2016; Glasze and Mattissek 2009). To illustrate this position: A farmhouse in traditional architecture practiced over centuries is considered 'essential', a prefabricated plastic bungalow 'accidental', a copper beech 'essential' in most parts of Central Europe (with the exception of extreme locations), a cypress pine definitely not. This view is synthesised according to the 'old European' unity of truth, good, and beauty: Only the 'true' 'landscape', shaped over centuries by the reciprocal shaping of culture and nature, can be 'aesthetically pleasing' and morally 'good'. Accordingly, 'accidental' elements are regarded as morally 'bad', aesthetically 'ugly' and ontologically 'unauthentic'. In this context, it is often referred to the high speed of change processes: "If the change takes place quickly, it initially appears disharmonic, since the balance is disturbed for a long time" (Lautensach 1973, pp. 26–27). Accordingly, the norm of the preservation of the 'essential landscape', often referred to as the 'historically grown cultural landscape' (e.g. in the case of Quasten 1997; Wöbse 1999, 2002), is derived. This is ultimately a naturalistic fallacy – also called 'Sein-Sollen-Fehlschluss' (Stuhlmann-Laeisz 1983) or 'Hume's Law' as Hume regarded the Is-ought problem (cf. Quante 2008, p. 122; see also: Hume 1978; Sen 1966) – because from the existing the desired is derived, i.e. in this context: Just because a landscape is existing in a certain way, it cannot be concluded that it should continue to exist in this way (Kühne 2019c). Although, of course, factual developments can be assessed as 'good', 'right', or 'better' in an evaluative or normative attitude, such an assessment then needs to be justified in an evaluative or normative way. Without a proven normative justification, the conclusion of factuality to normativity is, however, a naturalistic fallacy of the kind aforementioned. Especially in the contexts of (landscape) planning, including political and economic regional development, the essentialist understanding of landscape is still strongly present today, as it provides (apparently) comprehensible norms and can thus be used strategically against alternative interests (industrial development, settlement expansion, rational agriculture, etc.; Fürst et al. 2008; Gailing et al. 2006; Gailing 2012, 2015).

In landscape research in the positivist tradition of thought, landscape is not ascribed to an 'essence', but rather understood as a physical and real object. The research of this 'object landscape' is carried out on the basis of scientific research tradition by measuring, weighing, and counting, not only with regard to the objects of natural science but also social science research: "Quantification, mathematisation and computer modelling seemingly offered unlimited potential for unravelling the spatial fix of human affairs" (Tilley 1997, p. 9). Thus, positivist landscape research is based on the observation and abstraction of individual phenomena by "generalizing the collected observations inductively by the mind" (Eisel 2009, p. 18), whereby the intermediate step of spatial structuring of the data is often performed by a Geographical Information System (GIS). In order to improve the comparability and processing of data, an attempt is made to also subject genuinely qualitative data, such as aesthetic and emotional attention to space, to a quantitative evaluation scheme (among many: Frank et al. 2013; Hunziker and Kienast 1999; Kianicka et al. 2006). Based on descriptive data and their analytical synthesis, it is possible to derive predictions about possible states of landscapes interpreted as objects by means of models, but their normative content remains manageable; in the end, positivist research is subject to the principle of Freedom of Values (e.g. Albert 1960). Accordingly, it is difficult to transform the results of positivist research into political and planning norms. In planning or nature conservation, essentialist normative interpretations and assessments are frequently used, which are then applied to descriptive, analytical, or modelled 'underpinnings' (in this context: Hupke 2015; Kühne 2018h, 2019c).

If the constitutive level of landscape lies in the essentialist view in its 'essence' to be found behind materiality or in positivist research in its materiality, then the constitutive level of constructivist theories lies in the social (among many: Aschenbrand 2017; Fontaine 2017; Kühne 2018f, 2019c; Weber 2018b). Currently, three perspectives of constructivist landscape research can be identified (specifically in German landscape research): the social constructivist, the discourse-theoretical, and the autopoietic systems-theoretical. The social constructivist perspective relies on the phenomenological sociology of Alfred Schütz and the sociology of knowledge of Peter L. Berger and Thomas Luckmann (Berger and Luckmann 1966; Schütz 1960 [1932], 1971 [1962], 1971; Schütz and Luckmann 2003 [1975]) that 'landscape' is the result of social negotiation and conventionalisation processes. Landscape is thus viewed as physical spaces in accordance with socially educated, culturally differentiated patterns of interpretation and evaluation that are socialised for the individual (in Anglo-Saxon, see Greider and Garkovich 1994). An essential topic of social constructivist landscape research involves questions pertaining to the transmission of social patterns of interpretation and evaluation in socialisation processes (Kook 2008; Kost 2017; Kühne 2008a, 2014c, 2017, 2019b; Nissen 1998; Stotten 2013; see detailed Box 1), but also questions exploring the social distribution of power with regard to the implementation of aesthetic and normative ideas on landscape in particular (Cosgrove 1993; Cosgrove and Daniels 1988; Kühne 2008b, 2014d; Leibenath 2015; Mitchell 2002, 2008). This question focuses in particular on the discourse-theoretical perspective (among others: Leibenath and Otto 2012; Otto and Leibenath 2013; Roßmeier et al. 2018; Weber 2015, 2019a; Weber et al. 2017; in English context: Wylie 2015). Following on from the Discourse Theory of Ernesto Laclau and Chantal Mouffe (Laclau 1996; Laclau and Mouffe 1985; Mouffe 2000, 2005, 2007) the study examines which interpretations and valuations of landscape become hegemonic, which other interpretations exclude them. Central questions are 'which positions are connected and how', 'which are excluded' and not least 'how – as a result of which powerful constellations – a change of interpretation and evaluation can be carried out', an example being in the form of a shift from the interlinking of 'wind turbines are ugly' to 'wind turbines are sublime'. The autopoietic-system-theoretical perspective goes to the sociology of Niklas Luhmann (1984, 1986, 2017) who, for his part, particularly focused on the systems theory of Talcott Parson's (1991 [1951]) and radical constructivism (Glasersfeld 2000; Maturana 1985 [1982]; Varela et al. 1974). Like Parsons, he believes that society is divided into subsystems that take on specific tasks for society (such as the economy for the supply of goods and services, politics for fundamental decisions about the development of society). From the radical constructivists he adopts the position that the 'truth about himself and the world' cannot be grasped by man. Combining these two positions, he comes to the conclusion that social subsystems view the world only according to their specific codes, i.e. money (economy), power (politics), knowledge (science), etc. 'Landscape' is thus only dealt with in terms of the extent to which it can be used in the economy to generate (or lose) money (tourism, extraction of raw materials), power (politics; e.g. resistance to wind power), knowledge (e.g. problems of acceptance from economic activities), topicality (mass media; reporting on resistance to spatial change etc.), and to what extent it can be used in the economy to generate (or lose) money (tourism, extraction of raw materials), power (politics; e.g. resistance to wind power), knowledge (e.g. acceptance of problems from economic activities), current relevance (mass media; reporting on resistance to spatial change), etc. However, society has not differentiated any subsystem that deals with the environment or nature or landscape in its entirety, resulting in unilateral claims (Luhmann 1986). Thus, the concept of ecosystem services can be interpreted as an attempt to translate the 'value' of nature and landscape into the code of the economic system (Kirchhoff 2019; Kühne 2014b).

Box 1: The differentiated socialisation of landscape

The socialisation of landscape is not a continuous process, but a thrilling one (especially when biographical challenges related to landscape exist), and can in principle be divided into two phases, which can be followed by a third. In the first phase the 'native normal landscape' is formed, in the second phase the 'stereotypical landscape'. In the third phase, the optional introduction to 'landscape-related special knowledge stocks' will take place (Kühne 2019c; see also Kühne 2008a; Stotten 2013). The 'home normal landscape' is the result of the appropriation of the living environment in childhood, mediated by parents, if available by (especially older) siblings, grandparents, later also in the peer group, etc. The 'home normal landscape' is the result of the appropriation of the living environment in childhood. It is "filled with first memories of regional language, noises, smells, colours, gestures, moods and speaking things and deeply anchored in the memory" (Hüppauf 2007, p. 112) and has a strongly positive emotional connotation; an intensive norm of permanence is attached to the material objects it occupies (Kühne 2011; Kühne et al. 2013; Stotten 2019; Weber et al. 2019). The 'stereotypical landscape', on the other hand, is not based on unquestioned familiarity, but on socially shared norms, especially aesthetic norms (Kühne 2008a, b, d). It is based in particular on secondary information from school lessons (especially schoolbooks), films, illustrated books, the increasing importance of the Internet, etc. In addition to aesthetic norms, ideas of individual usefulness are also formed in this phase (e.g. as a backdrop for sporting activities). The 'stereotypical landscape' can be understood as a common sense understanding of landscape in a social association (e.g. a language community). 'Expert knowledge' shall be acquired during vocational training, in particular through a relevant scientific course of study. According to academic standards, they are strongly cognitive in character, but are also frequently subject to subject-specific deficit orientation (especially when essentialist patterns of interpretation are conveyed). The subject-specific 'expert special knowledge inventories' are even more unstable than the 'stereotypical landscape', since they are subject to competing paradigmatic fundamental orientations (such as successionism in demarcation to the preservation of 'historical cultural landscapes'; cf. Hokema 2015; Kühne 2008b; Leibenath 2014; Wojtkiewicz 2015).

The positivist approach, in particular, originates with a dichotomous separation of subject and object, the autopoietic-system-theoretical approach is solely related to society (material objects exist only as objects of communication), which also applies in a somewhat diminished way to discourse theory, since in the previous investigations the discursive negotiation of material spaces conceived as landscapes was examined in particular. Social constructivist landscape theory also considers material objects (especially when it is extended to include symbolic-interactionist or ethnomethodological elements), although it relies heavily on cognitive references. This is where the first of the 'intermediate' approaches comes in: phenomenology. It has the same roots as the writings of Alfred Schütz (for a comparison of the two approaches see Kühne 2019c, d), but the focus is different: first, it is concerned with the (aesthetic and emotional) experience of landscape, less with cognitive constructing; second, this experience is understood as an individual, which means that the aforementioned individual dimension (admittedly in the social interweaving) of landscape is thematised; third, fewer symbols move into the focus of interest than atmospheres (see among many: Kazig 2007, 2013; Nogué i Font 1993; Tilley 1997; Wylie 2005). Like phenomenology, actor-network theory (ANT) and assemblage theory strive to dissolve the subject-object dichotomy. The ANT, going back in particular to Bruno Latour and Michel Callon (Latour 1998, 2002 [1999]; Callon 1999), is anxious to break up the subject-object theory with the help of the network concept and thus to understand material objects (whether 'technical' or 'natural') with an independent social existence as 'actants' (Kneer 2009; Kühne 2019e). If the ANT endeavours to generate a new understanding of the world (with its own terminology), the assemblage theory endeavours to integrate materialities into constructivist approaches to the world, here landscape. Thus, material things become socially relevant when they are negotiated discursively (Mattissek and Wiertz 2014). This raises the central question of how material things (socially) can have a crucial effect? (van Wezemael and Loepfe 2009).

The genesis of the understanding of landscape (whether in relation to the common sense of the 'stereotypical landscape' or in the context of scientific discussions in the form of 'expert special knowledge') is of particular scientific interest when a constructivist perspective is pursued, since the social construction of the world in general and landscape in particular is the focus of interest. But it can also be advantageous from the perspective of intermediate landscape research, since it can trace indications of a changed relationship between society and materialities. Since for positivist landscape research landscape is understood unquestioningly as a given object, the question of the genesis of understanding is of secondary importance. The same applies to the essentialist perspective, in which language becomes interesting if it contains specific regional expressions (e.g. in the form of dialect). Otherwise, language here is a pure medium for describing the 'essence' of landscape. Particularly due to the specific appropriateness of constructivist perspectives on landscape, the present study is theoretically framed with these. Since this involves in a special way the linguistic expression of landscape, there is now a brief reflection on the meaning of language in the production of social 'realities'.

2.2   The Importance of Words and Language in the Creation of the World

Explanations of the 'history of concepts' of 'landscape' assume that the meaning of this special concept, and of concepts in general ('semantics'), is subject to a historical change. In the humanities and cultural sciences, 'conceptual history' is therefore brought into connection with 'historical semantics' (cf. e.g. Dutt 2011; Gumbrecht 2006; Koselleck 1979; Müller and Schmieder 2016). Müller and Schmieder (2016, p. 18) regard 'historical semantics' as a "linguistic part of cultural semantics" and heuristically as a "generic term for methodological approaches which, like the history of problems, ideas or concepts in the broadest sense, deal with diachronic linguistic changes [...]". If historical semantics represents the "context of conceptual history" (ibid.), conceptual history examines the origin and the change of meaning of concepts within a historical context. Against a logical understanding of concepts as "abstract objects" (Mittelstraß 2004, p. 270), which supposedly cannot be subject to any historical change, it is justified in view of the difficult question of whether and to what extent "new" concepts arise "in the strict sense" through linguistic changes to assume "that a concept develops or that there is a history of differentiation or history of a word's meaning. In this sense, every reconstruction is also an element" of a conceptual history (Mittelstraß 2004, p. 271). This logical understanding of concepts stands in a philosophical-scientific tradition that subordinates the forms, functions, and meanings of language to logic: "Natural and historical aspects of language, on the other hand, are regarded as accidental in the sense that they leave thinking, being and truth untouched, or in other words: *meaning remains pure* – so that pure science is possible" (Borsche 1996, p. 9; emphasis in the text). The core of this tradition is the Platonic-Aristotelian-inspired "stoic doctrine of the generality, immutability and incorporeality of meaning" of language, the "doctrine of the speech-free or general nature of thought" (Borsche 1996, p. 10) as well as the doctrine, speech, or language are "expressions of general thoughts" and thought is only the "grasping of natural facts [...], which it is scientifically by words or better still by numbers and symbols to truly represent" (Borsche 1996, p. 11). Language is understood, as it were, as a means of transporting object-like, thus finding immaterial and unchangeable meanings in a 'world', which are to be truly taken up by a 'language-free reason' (Borsche 1996, p. 11) and mediated for an appropriate knowledge of the world. This "Leibniz Programme" (Mittelstraß 1998, p. 35) therefore understands knowledge in general and scientific knowledge in particular as "propositional knowledge" (Mittelstraß 1998, p. 47), i.e. as a system of scientifically 'true' statements ('propositions'). This view is therefore referred to as 'propositionalism' (cf. Gabriel 2005).

This view only changed when language was no longer clamped into a subject-object difference as a medium of meaning, in which language truly represents or depicts a 'world' for recognizing subjects and merely transports it. Language, on the other hand, was located in the Twentieth century "in the field of tension between different subjects who form themselves as individuals and between whom [...] only that which one tends to call 'world' in the philosophical sense of the word and what Humboldt calls world views in order to express the inescapability of their perspective diversity can arise" (cf. Borsche 1996, p. 11). According to Humboldt, "Language" is "the means by which man simultaneously forms himself and the world, or rather becomes aware of himself by separating a world from himself" (Seidel 1962, vol. 2, p. 207; quoted in Borsche 1996, p. 11). Linguistic meanings are not preformed in a platonic realm of ideas and naturally-objectively predetermined in every communication, but they emerge and pass away in and through communication and consolidate or perpetuate their content, for example "through convention or institution [...]; concepts live, they live in the mutual recognition and community of those who use them in their words" (Borsche 1996, p. 12). Language is therefore a central mechanism of the mediation of an objectified social world. As a point of departure and reference for the everyday world, it provides categorizations, typifications and anonymizations. Thus, language is not an instrument for depicting reality, it has a performative effect, it produces realities, it structures perceptions and thus represents a social "system of signs and rules" (Werlen and Weingarten 2005, p. 192) which is mostly based on unreflected typings (Schütz and Luckmann 2003 [1975]). This system is characterised by a certain variability, but attributions of meaning are "not arbitrary, but contingent, i.e. they are determined within the coordinates of a culturally (spatially) and historically (temporally) specific statement system" (Wachholz 2005, p. 106).

The world-constituting role of language thus also affects the sciences and their forms of communication. Everyday scientific life consists of research, and research "is essentially action, practice, under conditions of theory and methods, but not itself as theory and method. This is research only if it presents itself in its results or describes the paths it has taken" (Mittelstraß 1998, p. 46). The rationality of action and that of knowledge, 'doing' and 'knowledge' are equally intertwined and "related to each other (...). Our

idea of science therefore falls short if it only grasps its theoretical, 'propositional' sides"
(Mittelstraß 1998, p. 47). Science is not only a system of scientifically 'true' statements
or 'propositions' (cf. Gabriel 2005), but also a system of acting and interacting scientists
who determine the criteria of scientificness, scientific 'truth' and corresponding criteria
of historically achieved and provisionally proven knowledge within the framework of
specific purposes, goals, standards, and customs. There is no 'way to the truth', however
given, which the sciences would have to commit and from which a 'wrong' choice of
theory would be the result. Scientific terms – this applies a fortiori to terms of pre-scientific colloquial language – are also the 'living' product of the 'mutual recognition and
community' of the scientists (in this case).

From a critical perspective, however, the function of language in social contexts –
here in relation to landscape – can also be questioned. Language is – according to Burckhardt
(2004) – "an instrument for implementing action; [for] all rule is linguistic." Persons
without landscape-related secondary socialisation often remain isolated in silence on
topics of landscape design and development, since they only have the choice "between
the 'official language' they have been forced to use and their own colloquial language"
(Bourdieu 2005 [1983], p. 27) if they want to communicate with professionals about
landscapes – especially in legally prescribed or voluntary participation procedures, for
example in urban land-use planning or the designation of protected areas – they only
have the choice "between the 'official language' that is foreign and imposed on them and
their own colloquial language" (Bourdieu 2005 [1983], p. 27). The silence of those who
are not professionally concerned with landscape is therefore often due to a strategy of
avoidance: The self-confidence and the technically precise and thus linguistically dex-
terous expression of the professionals – whereby this language in the sense of Gelfert
(2000, p. 85) is described as "intimidating kitsch" by irrational myths and high-sounding
forms of expression – they are intimidated and refrain from expressing their interest.

In the following, we will first deal with the 'everyday world' understanding of landscape before we will deal with the terminological version of landscape in (different) scientific disciplines of the German-speaking world.

The Genesis of the Landscape Concept in the German Language Regions – The Common Sense Understanding

After preparatory work from different disciplines with historically significant anthologies (Hartlieb von Wallthor and Quirin 1977; Ritter 1975 [1953], 1996) which (also) deal with 'landscape', a cultural-historical reconstruction of the *genesis* of the German (and partly also European) concept of landscape has meanwhile been established, which diachronically distinguishes three main meanings from 'landscape (see e.g. Hard 1991, 2002a; Jackson 1984; Jessel 2005; Kühne 2018f; Leibenath and Gailing 2012; Schenk 2002, 2017; Haber 2006; Kirchhoff 2017):

1. an originally legal-territorial-political concept of landscape: landscape as a *regio* (Hard 1977; Kortländer 1977),
2. a later aesthetic-emotional concept of landscape: landscape as 'image' and 'soul symbol' (German: *Seelensymbol*; Gruenter 1975a [1953]; Hard 1977),
3. a presently established physical concept of landscape: landscape as 'earth space' (Oppel 1884) or 'earth region' (Leibenath and Gailing 2012).

The semantic transitions between the main meanings can be described or reconstructed as 'secondary formation' (Schenk 2013) in the sense of a reinterpretation of meaning from the respective preceding meaning or also as a 'format jump' between different aesthetic design formats (Hauck 2014) and imply conceptual connotations that have influenced the understanding of landscape in general and geography in particular. These are, firstly, the 'regionalizing' connotations of landscape as a region, which promoted thinking in earth spaces and divisions, and, secondly, the 'physiognomic' connotations of the *landscape eye* of landscape painting, which created holistic and aestheticizing accesses to landscape (cf. Hard 1977, p. 15; Schenk 2011, p. 12).

This history of concepts, including the reinterpretations of meaning, can hardly be understood unless a brief look is taken at a specific prehistory. This prehistory is to be reconstructed from texts and testimonies reaching far into the past, some of them classical, which thematise and describe spatial phenomena that we are prepared to subsume today under the term 'landscape', although such a subsequent attribution to these phenomena is often historically or factually inappropriate. The semantic bridge between this prehistory, the history of concepts, and the current discussions in German-speaking countries consists of an allegorical fantasy of ancient poets and painters that revolves around the motif *arcadia*.

3.1   On the Classical Prehistory of the Concept of Landscape – Origins Beyond the German-Speaking World

In recent discussions about concepts, theories and concepts of 'landscape', the term 'Arcadian associations' also plays an important role (cf. e.g. Eisel and Körner 2009; Hard 1991; Hokema 2009, 2013; cf. also Kühne 2018d; Prominski 2004). What is meant is the fact that the concept of landscape refers to traditional ideas of a 'beautiful' or 'lovely' as well as 'natural' and 'harmonious' scenery (Jackson 1984). Historical origins of this paradigm of imagination are the literary-rhetorical topoi *arcadia* and *locus amoenus*. Before the depiction and transfiguration in poetry and fine arts, the landscape commonly associated with the *arcadia* motif was the product and appearance of a 'land use system' of the time (Küster 2012a) in its economic interdependence with the urban environment (Frizell 2009). Erwin Panofsky, in his famous essay 'Et in Arcadio ego' (1975 [1936], pp. 351–377) can show that the Arcadia motif was originally understood as a humanistically turned *memento mori*, but with Nicolas Poussin took on a different meaning: the evocation of a past 'golden age'. Previously Vergil had already seen the tragedy, which in Theokrit still appears 'as unyielding reality', "through the mild veil of anticipatory or retrospective sensation" and he changed "mythical truth in elegiac feeling" (Panofsky 1975 [1936], p. 356). Bruno Snell (1975 [1945], pp. 257–274) continues here and shows that Vergil is the poetic discoverer of Arcadia as the "land of shepherds, the land of love and poetry" (Snell 1975 [1945], p. 257). While still "Theokrit had described the shepherds of his homeland realistically and ironically in their everyday milieu, Vergil saw an elevated and transfigured existence in the life of theokritic shepherds" (Snell 1975 [1945], p. 258): Everything now stands in the 'glimmer of feeling' and the 'sentiment', *arcadia* becomes the "land of gilded everyday life" in a "world of sentiment" (ibid., p. 263). Ernst Robert Curtius (1954 [1948], pp. 191–209) describes in particular the emergence and effectiveness of literary 'ideal landscapes' such as 'Arcadia', the 'Hain' and the 'locus amoenus'. Its "minimum components consists of a tree (or several trees), a meadow and a spring or stream. Birdsong and flowers can be added" (Curtius 1954 [1948], p. 202). Rainer Gruenter (1975a [1953]) criticised Curtius in this point because he conceived of the 'locus amoenus' as a landscape and thereby referenced the "application of a term inspired by modernism [...], which corresponds to a completely different expressive-historical behavior" (Gruenter 1975a [1953], p. 200) "to the entire historical space of her field of research" (Gruenter 1975a [1953], p. 199), i.e. the modern concept of landscape to the pre-modern literary topos of the 'locus amoenus'. This confusion of graceful garden-like nature scenery and landscape can still be found today among nature conservationists (cf. Trepl 2012, pp. 215–238).

Greek poetry knew neither the term nor the "thing" landscape, which "was unknown to the Greeks, at least in its early days" (Elliger 1975, p. 2). The terms *tópos* (place) and *chõros* (space, region) were widespread. The same applies to the visual arts (ibid., p. 4), in which 'landscape' is represented only as an iconising 'abbreviation' (ibid., p. 1), without viewing the individual natural phenomena together to form a spatial context in today's sense of 'landscape'. In literature, this can be seen in passages from Homer's *Odyssey* (Homer 1994, 6. song, pp. 290–294; 9. song, pp. 132–135) and Theokrit's *poems* (Theokrit 1999, VII. poem, pp. 133–145) as well as in the philosophical texts of Plato and Aristotle. In the well-known scene in the Platonic dialogue of Phaidros, Socrates speaks of the fact that *chõría* (spaces) and *déndra* (trees) cannot teach him anything (Burnet 1962, 230d). Aristotle does not know such a word either; in *politics* (Aristotle 2009, pp. 305–306) he speaks only of the chõros (the space surrounding the city; translated as 'land') as a source of supply for the city (pólis). The same can be said for 'landscapes' in *roman poems*, which, as in Catull and Vergil, only acquire significance as mirrors of "the mental condition of the main character" (Elliger 1975, p. 434) oriented on individual 'scenic' phenomena and for which Latin expressions such as 'regionis', 'terra' or 'villa' (as in Plinius 1990 [first around turn of the first to the second century]) are used. The famous villa letters of Plinius (1990) are significant in terms of their historical impact because they are a "villa with a view" (Küster 2009, p. 11) from which one could see the garden and the mountain ranges of the Appenine. These letters point ahead to the famous 'view' of Petrarca from Mont Ventoux to a 'landscape' or, accordingly, from Boccaccio in the Decamerone (cf. Küster 2009, pp. 8–14).

3.2 Medieval Roots in the German-Speaking World

The 'original' meaning of 'landscape' (Kortländer 1977, p. 36) is *regio* or *provincia*, i.e. region, territory, indigenous population, "settlement unit" or "political unit" (Hard 1977, p. 13) and is etymologically derived from *lantschaft* (Gruenter 1975a [1953]) or *lantscaf* (Müller 1977). The word *lantscaf* is first used in the ninth century (Gruenter 1975a [1953]). The word segment 'land' in 'landscape' as a translation of 'region' (Haber 2007, p. 78) has different meanings: as a contrast to a water body, as a territorial designation (Germany) or as a "surface for building, digging or depositing". But "only 'land as contrast to the city' (from which also the property word 'rural' is derived) [has] a relation to 'landscape'" (Haber 2007, p. 78; emphasis: K.B. and O.K.). The word segment "shaft" is a verbal abstract derived from *skapjan* 'to create' (Müller 1977, p. 4), which, in addition to abstracta (master-hood) and collectiva (man-hood), also includes spatial designations (Müller 1977) which refer both to the meaning of something belonging together (neighbour-hood) and to the meaning of what has been created through human work and design (Haber 2007, p. 78). Thus existing in the Old High German etymology both the already mentioned, always resonating holistic character of "landscape" as either a spatial structure or as an aesthetic figure (cf. Hard 1977, p. 15; Schenk 2011, p. 12) as well as the aspect of working, creating, inventing and shaping aestetically (*shape*; Haber 2007, p. 78; cf. Termeer 2007) are connected to each other.

The original meaning as *regio* comprises 1) a spatial-legal settlement, 2) a territorial-political area and 3) a person-collective variant of meaning. 1) The primary spatial-legal settlement meaning of 'regio' encompasses territorial, ethno-social, and legal-political dimensions (Stiens 2009, p. 38; Tesdorpf 1984, p. 43), especially "social norms in a country" (Müller 1977, p. 7) and a "time-honored local-regional customary law" (Trepl 2012, p. 161) of "different groups involved in the rule of the land" (Trepl 2012, p. 163). The 'social' (Kühne 2018f, p. 32) and legal connotations were also applied in the following centuries to the "country in which such standards are valid" (cf. Groth and Wilson 2005; Müller 1977, p. 7). Landscape as a "larger settlement area with certain historical-social similarities" is therefore defined by "regional requirement, regional custom" (Müller 1977, p. 7) and in the early Middle Ages not yet defined by exact spatial boundaries or physical-geographical units, but rather as a 'social' component of meaning. 2) Only in the following centuries did the "ruling functions" of a sovereign in a territory or a province become "regionally located and demarcated from each other" (Kühne 2018f, p. 32); a function that still persists today in territorial designations, such as the name of the Swiss canton 'Basel-Landschaft'. One can also speak of a space defined by legal policy, which, however, is "within a larger whole (sovereign state)" (Müller 1977, p. 7) as located here. 3) At the end of the eleventh century, the spatial significance component of a "strip of land" is also transferred to the "politically capable", to the "estates" as "representatives" of the "whole landscape" (Hard 1977, p. 14). This component of meaning can still be found today in expressions such as "*Landstände*" [Country Estates], "*Ständeversammlung*" [Estate Council] or "*Ostfriesische Landschaft*" [East Frisian Landscape; Müller 1977, p. 8). In the late Middle Ages 'landscape' became 'a precise expression of human laws and legal institutions' (Kühne 2018f, p. 32; cf. Olwig 1996).

Finally, this meaning as a 'collective of persons' is successively transferred to the political or natural space populated by these groups of people. 'Landscape' can now be used "also to describe natural units" (Müller 1977, p. 9) and can be equated with 'region' and the Latin *tractus* ('region') in the sense of smaller and manageable units of space. This "early led to aesthetically motivated formulations", such as "beautiful area" near Herbort von Fritzlar (Müller 1977). Thus, a first basis has been laid for understanding the real earth surface in its geographical-ecological structure with landscape from the seventeenth century onwards (Tesdorpf 1984, p. 42).

An *aesthetic* attention to place and space, which in ancient poetry and painting was given within the framework of Arcadian topoi, but with the exclusion of areas not made by humans was (cf. Appleton 1975), in the Middle Ages, replaced by a secular and transcendent view and largely limited to a function oriented beyond within a closed religious and feudal world view: "The gaze is directed to heaven, but not to its earthly manifestation" (Lehmann 1968, p. 9). In the Middle Ages, therefore, there was no aesthetic turn to landscape as an *independent* subject in painting and poetry. In the early Middle Ages, artistic depictions of nature in landscape painting were predominantly under the spell of religious motifs and corresponding allegorical depictions of salvation history (Böheim 1930; Büttner 2006; Lehmann 1968; Steingräber 1985; Assunto 1963; Erb 1997). In this religious context, for example, medieval painters were obliged to "express the divine in works of fine art" (Büttner 2006, p. 36). Medieval 'artists' (lat. *artifex*) did not produce any subjective-creative works of art, but their "works of art were an essential part of religious life in that age. The representation of Christ, the Mother of God, the construction of a cathedral were not aesthetic tasks, but deeds to the glory of God, worship" (Grassi 1980, p. 98). Such an artist was an instrument of divine revelation and his works "did not arise from the endeavour to objectify a colourful world of being in its diversity and its richness of relationships, but to depict the salvation history of mankind and the symbols of its salvation" (Böheim 1930, p. 82; cf. also Trepl 2012). 'Secular representations' in the late Middle Ages to the early modern period served as profane feudal or merchant bourgeois demonstrations of power and possession (Schneider 2009, p. 15) in book illuminations (Steingräber 1985; Warnke 1992) or frescoes (Steingräber 1985; Eberle 1980; Lützeler 1950). The original spatial-territorial connotations of the word 'landscape' were transferred into the painted 'landscapes', so that in such representations "nothing other than the owner's pride and limitation of the landowner" (Warnke 1992, p. 65) can be found.

Authors like E.R. Curtius (1954 [1948]) and R. Gruenter (1975a [1953]) were able to show how landscape in German and European literature of the Middle Ages has not yet been interpreted in today's sense as a whole of perception (cf. Steingräber 1985; Tesdorpf 1984; Schneider 2009), but analogous to the *Abbreviaturen* (Engl: abbreviations; Büttner 2006, p. 33) was represented in landscape painting as a "stenogrammatic nature note". The principle of *Topothesie* (English: topothesis), already developed in ancient poetry and rhetoric, is addressed (Büttner 2006, p. 37; Schneider 2009, p. 15), by means of which landscape elements that have been handed down abstractly from literary *tópoi* are deduced and put together like set pieces into a nature scene and presented as 'locality' (topo-thesie). In particular, the ancient literary and allegorical topos of the *locus amoenus* (Curtius 1954 [1948]; to the criticism of Curtius: Gruenter 1975a [1953]) perpetuated. This can be exemplarily observed in the classical poems *Parzival* by Wolfram von Eschenbach (Eschenbach 1998 [first around 1200/1210]), *Tristan und Isolde* by Gottfried von Straßburg (Straßburg 1991; cf. Gruenter 1975b [1953], pp. 307–312), *Carmina Burana* (1974) and the *rose novel* by Guillaume de Lorris and Jean de Meun (Lorris and Meun 1976–1979; cf. Hennebo 1987, p. 66). The *Capitulare de villis vel curtis imperii* by Charlemagne (Brühl 1971) shows how landscape can still be absorbed into agriculture. Albertus Magnus' *De vegetabilibus* (Magnus 1867 [first around second half of the thirteenth century]) shows how the literary topos of the 'locus amoenus' is to become reality in the form of a 'pleasure garden'. Albertus' sketch of a clearly structured garden can also be classified "as a preview of a later epoch of garden art", "in which the garden is formed and its elements arranged" (Hennebo 1987, p. 37).

3.3   The European Influence I: Aestheticisation in the Renaissance

To see nature aesthetically as a landscape required centuries of development before the 'landscape eye' necessary for an aesthetic landscape view could be achieved (Riehl 1996) and formed accordingly. This history of mediation essentially consists of three stations: 1) The visual arts of the *Renaissance,* which represent idealized *picturesque* images of nature and develop the technical term of 'landscape' for it; 2) literature, which takes up such picturesque images of nature and in *Romanticism* transfers landscape aesthetically-emotionally into the interior of the sensitive subject and transfigures it into a 'soul symbol' and a 'mood landscape'; 3) the English landscape garden, which projects the established sight pattern 'landscape' conveyed by painting and literature into a physical space as a kind of 'walk-in image', transforms it into design and thus reifies it as supposed 'reality'.

It was only from the sixteenth century onwards that 'landscape' in Europe was also referred to as an 'aesthetically and emotionally conceived image of nature' (Hard 1970c, p. 34). In the late medieval panel painting of the fiftheenth and sixteenth centuries in southwest Germany, "landscape" was initially regarded as the "technical term for the painting depicting a section of nature" (cf. Gruenter 1975a [1953]; Müller 1977, p. 9). This 'technique' required some preconditions. The "two-dimensional object to object relationship" (Gruenter 1975a [1953], p. 200) of natural objects as a "conglomerate of isolatable parts" (Friedländer 1947, p. 19–20) in an 'aggregate room' (Panofsky 1980 [1927]), i.e. the early medieval 'abbreviations' (Büttner 2006, p. 33) as a "stenogrammatic nature note" (Gruenter 1975a [1953], p. 200) as part of a *topothesis* already developed in antiquity (Büttner 2006, p. 37; Schneider 2009, p. 15) first of all had to be classified in a "spatially logical context" (Friedländer 1947, p. 19) and overcome in favour of a 'system space' (Panofsky 1980 [1927]), in which the 'meaning-perspective surface' became a space constituted from a subjective 'point of view' (Friedländer 1947, p. 19) from a 'central perspective' (Gruenter 1975a [1953]). Only then can a "conglomerate of natural objects (...) be seen as a landscape" (Gruenter 1975a [1953], p. 200). Landscape is now a "picture" or a "visual figure" (Hard 1991, p. 14) in the sense of a whole view, later also a symbolic 'sense-image' (Kortländer 1977; Kühne 2018f; Kirchhoff 2017; Schneider 2009) for ideas (Berr 2008). 'Landscape' as a technical term in painting has then successively established itself in general European language use in the national variations (*landschap, paesaggio, paysage, landscape;* Gruenter 1975a [1953], p. 198). The Dutch and English landscape painting of the late sixteenth and early seventeenth centuries and the Dutch landscape painting of the seventeenth century show a line of development up to the Romantic period.

Although the landscape painting that developed in the Renaissance was already 'closer to reality' than that of the Middle Ages due to the avoidance of allegorical representations, there can hardly be any talk of 'naturalism' in the modern sense. The central perspective developed during the Renaissance and the exploration of the interplay of colours and shades enabled a more detailed representation of physical-spatial phenomena and moods (Cosgrove 1984, 1985; Eisel 1982; Piepmeier 1980). Nevertheless, the art of the Renaissance was still characterised by idealizations in the opposite direction. These 'technical' innovations and idealizations were made possible by the fact that the individual increasingly freed himself from religious convictions and compulsions and art freed itself from sacred purposes within the framework of a world view now oriented towards this world. People discovered themselves and the *worldly* world. 'Reality' now appears as a 'world' or 'nature' which is harmoniously structured and ordered according to the laws of mathematical principles, which can be experienced empirically and analysed mathematically. 'Beauty' is ultimately based on the "harmony (*concinnitá*) and proportionality (*proporzionalitá*) of all visible qualities, forms and colours" (Hauskeller 1995, p. 104). 'Beauty' in this world view represents the essential structure of 'reality' in which an essentialist understanding of landscape becomes clear. Art therefore has the task of exploring and discovering the laws of phenomena and the structural laws of reality. Art becomes quasi 'applied science' (cf. Plumpe 1993): for example, the laws of proportion, the laws of perspective, and the laws of colour theory are researched (see also Hauskeller 2005, pp. 27–32). Insofar as the arts bring to light the laws of reality as beauty, since the Renaissance one already speaks of "fine arts" (cf. Kristeller 1980, pp. 164–206). The artist was understood as a medium, as mediator, with Leonardo da Vinci as 'interpreter' between nature and man (cited from Hauskeller 2005, p. 31). In antiquity and the Middle Ages, however, he was a mediator between the world and God. The ancient and medieval mere *mímesis* (cf. Blumenberg 1957) was now replaced by a creative, perfecting *mímesis* as, so to speak, 'second nature', which had to idealise the shortcomings and imperfections of 'first nature' by bringing to perfection what nature itself had not been able to realise (Hauskeller 2005, p. 31). The necessary criterion for this 'presumption' – literally as well as figuratively – was the conviction of the artists as 'scientists' that for everything that can be found in the world there exists a perfect objective form. This is to be discovered through the observation and research of nature. To represent these perfections is the task of the artist. In this respect the latter completes the process of creation and thus becomes the 'alter deus' (Hauskeller 2005, pp. 31–32).

In landscape painting, against this cultural and art historical background, the type of idealised landscape developed "which was created in European painting of the seventeenth century and was given its [...] binding form by Claude Lorrain" (Riedel 1989, p 45; Conzen 2010; Cosgrove 1984, 1993; Kirchhoff and Trepl 2009a; Olwig 2007; Roters 1995). Lorrain created paintings that were consistently called "art landscapes, painted in the studio and enriched with scenes from mythology" (Spanier 2008, p. 282; cf. Schneider 2009, pp. 131–135). These paintings owe a large part of their effect not only to their respective content, but also to the novel consideration of light variations and "atmospheric phenomena", which show an "affinity to physical-optical theories of his time; to a certain extent they even anticipate them" (Schneider 2009, p. 134). Lorrain thus proved to be a child of his time by practicing art as an 'applied science'. This also includes the idealisation as an artistic-creative perfection of the depicted. In the typical endeavour of that time to continue ancient traditions, "the journey to Italy had become an integral part of the education of North Alpine artists" (Büttner 2006, p. 125; Howard 2011). The artist's work is not a composition, which is reflected in the depiction of Arcadian ideal landscapes, the less realistic renderings of precise sections of space, but rather were composed or "pieced together in the studio according to ideal patterns" (Burckhardt 2006a, p. 116). Here again – as Olwig (2008) describes – the creative process of 'creating' becomes clear: painters created (painted) land. According to the Arcadian *tópos*, it is about the "demonstration of a harmonious and peaceful nature", individual elements are "following the principle of 'electio', integrated into the composite landscape" and "form the repertoire of classical-ideal landscape painting" (Schneider 2009, p. 134). Landscape is depicted creatively idealizing in perfect harmony. With this painterly composition of Arcadian landscapes, a visual expectation geared towards visual stereotypes arose; ultimately, the visual arts were "pacemakers of our seeing and our experience of the landscape" (Lehmann 1968, p. 7).

Such visual expectations were subsequently transferred to physical spaces as well. The view pattern 'landscape' then functions as a guideline for the synopsis of objects in physical space (Cosgrove 1984, 1993; Haber 2007; Howard 2011; Hugill 1995; Schenk 2013, 2017). However, the transition from the ideal landscapes of the landscape painters to a landscape seen in physical space took place step by step. One example being the so-called Claude glasses (named after Claude Lorrain): the viewer moved these mirrors, tinted in the colours of Lorrain's pictures, back and forth "until one had captured a 'Lorrain landscape' framed, as it were, with the viewer characteristically turning his back on the landscape" (Groth and Wilson 2005; Kortländer 1977, p. 37; Löfgren 2002; Schenk 2013; cf. Shepard 1967; Foxley 2010). A further step in evolution is the theming of 'landscape' in *literature*.

After centuries of mediation through landscape painting (Büttner 2006, 2019; Gruenter 1975a [1953]; Hard 1977; Steingräber 1985; Schneider 2009) the aesthetic concept of landscape penetrated the 'fine literature' in the late eighteenth and early nineteenth centuries (Kortländer 1977) and thereby "into the general language of the educated" (Hard 1977, p. 14). The literary attention to the subject 'landscape' was ultimately carried by the endeavour "to see reality through the eyes of the painter and thus also nature through the eyes of the landscape painter; under the influence of such viewing habits, in […] 'elevated levels of language and style' […] the term 'landscape' was also transferred to reality, the 'picturesque aspect of nature'" (Hard 1977, p. 1). Thus, in literature it also happened that "the section of reality (and not only its artistic image) perceived with the eyes of the landscape was called *landscape*" (ibid.). Literature or poetry therefore required the preparatory work of landscape painting: *"Through the word, the phenomenon of painting has also penetrated poetry.* In the history of giving and taking between poetry and painting, the *landscape* is to be entered on the side of painting" (Gruenter 1975a [1953], pp. 203–204; highlights in the original). In other words: "poetic *landscape description* (…) begins with the literary discovery of *landscape painting*" (Gruenter 1975a [1953]); emphasis in the original). Admittedly, this is only *one* side of the problem, it is self-evident that poetry about landscape draws on a traditional treasure trove of forms from ancient and medieval poetics, especially the classical *locus amoenus* (Curtius 1954 [1948]). This means, "that the poet, even where he learned to see the landscape through landscape painting, remains under the spell of a topic well into the eighteenth century, which draws at least thematic boundaries to the poetic transposition of painting impressions" (Gruenter 1975a [1953], p. 206). In the course of the eighteenth century, for example, natural poets such as Barthold Hinrich Brockes (1680–1747), Albrecht von Haller (1708–1777), Salomon Geßner (1730–1788), Johann Heinrich Voss (1751–1826), and Gottlieb Klopstock (1724–1803) conveyed a sensitive ideal of nature in the tradition of idyll poetry in the German-speaking world. Such literary and poetic descriptions of nature were not only about a lovely Arcadian landscape. At the same time, since the eighteenth century, the aesthetic-literary appropriation of fear-and-scare-influencing nature or landscape has taken place, theoretically framed by theories of the sublime (cf. exemplary Kant 1993). In addition to poems and fiction, travelogues were also written and in this way conventions of experience and perception were also created, often with spatial concretisation, such as in the Alps, the Middle Rhine Valley, Dartmoor, and others (cf. Löfgren 2002; Schwarzer 2014).

3.4 The European Influence II: Between Idealisation and Reification of the Landscape – The Landscape Garden

After the baroque gardens in the seventeenth century, garden art in the eighteenth century staged the landscape garden as a counter-programme. The baroque garden with its geometric forms in layout and elevation plan symbolises not only man's claim to dominion over nature, but at the same time "the relationship of the (then) controllable to the unrestrained, to the territory of adventure and hunting" (Burckhardt 2006b, p. 50) and could therefore be understood as a symbol for the corporately organised society of absolutism. The landscape garden, on the other hand, was associated with the idea of freedom (Spanier 2008; Bender 1982; cf. also Dinnebier 1997) and one could therefore – in the sense of the Enlightenment – become the symbol of "a better society in the future" (Burckhardt 2006b, p. 51) in which man frees himself from the shackles of the absolutist order and recalls his natural fundamental rights (Olwig 1995). In this way, the baroque garden symbolises the urge to a "corrective of nature" (Seel 1996, p. 277) whereas the landscape garden opposes the idealization of nature, which is, of course, based on a culturalisation, namely a targeted arrangement of typical physical elements (Hasse 1993; Hugill 1995; cf. Daniels 1988 and 1999). It is a high-grade staging of nature (cf. Baumüller et al. 1997), as it were – 'walk-in pictures' in the garden space. In 1752 the poet Joseph Spence (1699–1788) described the landscape garden as a 'picture gallery' (Spence 1966, No. 1134) in the open air. Now the patterns of perception and expectations developed in landscape painting and landscape literature have been projected into the reality of the gardens. In particular, the strongly idealised pictures of Claude Lorrain (1600–1682), Nicolas Poussin (1593–1665), Salvator Rosa (1615–1673) and Jan van Ruysdael (1628–1682), semantically charged with Arcadian motifs, served the garden designers partly as direct models for their designs.

Centre of this enlightening critique of systems and culture was the notion of a "liberated", "paradisiacal" or "untouched" nature, in whose name the rationalistic mathematisation of nature and its fusion with the claims to power and domination of absolutism were denounced. For Shaftesbury, the mathematical-architectural principle of the French baroque gardens meant a "rape" of nature and the individual (Buttlar 1989, p. 12); Joseph Addison fought for the liberation of the subjective imagination from external constraints (cf. Buttlar 1989, p. 24) and for the liberation of nature from the formal-geometric design compulsion of the baroque garden principle (cf. Buttlar 1989, p. 12). The garden of the English poet Alexander Pope (1688–1744) in artistic style on his estate in Twickenham is considered "the starting point of the English landscape style" (Buttlar 1989, p. 25). One of Pope's mottoes was: "All gardening is Landscape painting" (handed down and documented in: Spence 1966, No. 606). New in this garden (cf. Buttlar 1989, pp. 25–29) was not only the free unfolding of the plants and the informal arrangement of spaces and nature motifs, but also the variety of garden scenes and views. For example, the view of the Thames from a grotto in the basement of the villa offered the viewer a picture that had previously only been known from landscape paintings. The media character of the three-dimensional picture arrangement in the landscape garden also led to the construction of two-dimensional backdrops, the so-called 'eyecatcher', which only had to create the illusion of a real building as a conveyer of mood or meaning. Lord Burlington (1695–1753), for example (cf. Buttlar 1989, pp. 29–36) from 1725 on created in Chiswick a garden of a new style, with vedutas arranged at the edge, showing views of buildings which, so to speak, present the picture programme of a walk through the garden's 'Picture Gallery'. On display are pavilions, gates, obelisks, columns, statues, a pantheon temple, a bathhouse, a grotto, an orangery, and views of the villa that Burlington had rebuilt on the model of the Italian architect Palladio, who worked two hundred years earlier.

The English landscape garden thus stands between an idealizing landscape painting and the reification of a spatial environment into 'landscape'. By allowing the landscape garden to be interpreted as a physical manifestation of the longing for the harmony of man and nature (cf. Spirn 1998), it has effectively historically shaped "the common image of a beautiful, harmonious and seemingly natural landscape" (Jessel 2005, p. 580) or an "ideal human-nature harmony" (Hard 1991, p. 14). Thus visitors and viewers believed, and believe, that they see 'natural' images or even 'images' of nature in the landscape garden, but in fact they see the creatively constructed image of an understanding of nature that is located between the longing for the *prehistoric* garden of paradise – as John Milton, for example, poetically described it in his "Paradise Lost" – and the *utopian* illusion of a *restored* harmony between man and nature (Buttlar 1989, p. 17–20).

The climax of the physical manifestation of the will to transform the physical space understood as landscape into an English landscape garden was reached in the second half of the eighteenth century. Here in the German-speaking world Franz von Anhalt-Dessau (1764–1800) endeavoured to transform his entire principality into a 'garden realm' (Haber 2005), with the aim of achieving a unity of aesthetics and economy based on the English model (Hirsch 1995). The term "garden empire" already illustrates the claim to territoriality of an ideal: "The concept of landscape to be realised there was based on the arbitrary designability of a landscape section in absolutism" (Franzen 2003, p. 122). Schneider (1989, p. 14) notes a change in ideology in the evolutionary phase from baroque to landscape gardening: "Whereas the owner of the estate was previously a ruler and ideologist in a single person who commissioned the gardener as a craftsman, in the romantic landscape garden the garden artist becomes the representative of the estate, ideologist and also ruler. This applies in particular if the garden artist or landscape designer acts in the name of the 'common good'" (Schneider 1989, p. 21). The purchase of land associated with the establishment of landscape parks meant "for the rural population the destruction of their basis of production, namely the arable and green areas, the expulsion of small farming families" (Schneider 1989, p. 21; cf. also Kühne 2008b).

Box 2: Goethe's turning away from landscaped gardens

In 1776 Duke Karl August of Thuringia (1757–1828), a friend of Franz von Anhalt-Dessau, left Johann Wolfgang Goethe a garden area with a garden house on the banks of the Ilm near Weimar. The reason for Goethe's own planning and gardening was the act of desperation by Christel von Lassberg on 16 January 1778, who committed suicide in the Ilm with Goethe's 'Werther' in her hand out of world-weariness. In memory of her, he built the famous rock staircase as a monument (Buttlar 1989, p. 161). He also built the *Luisenkloster* as a presentation of a ruin and participated in the design of the site to a park in the English style. At the end of 1777 Goethe had written the dramatic play *Triumph der Empfindsamkeit* (English: 'Triumph of Sensitivity'; Goethe 1999), in which he had already written the 'Age of Sensitivity' (Krüger 1972), which he himself helped to create, and a "superficial sentiment" associated with it (Buttlar 1989, p. 160), with irony and mockery. After his trip to Italy in 1786, his dissociative attitude became stronger, not only with regard to the sensitive emotional enthusiasm, but also with regard to the idea of a landscape garden in general. His participation in the further design of the park was henceforth limited to architectural contributions such as the 'Roman House'. Together with Friedrich Schiller and Heinrich Meyer he wrote the fragmentary sketches *Über den Dilettantismus* ('About Dilettantism'; Goethe et al. 1896) in 1799, in which he considered the mixture of nature and art as well as the lack of austerity of form to be a great disadvantage of garden art (cf. Buttlar 1989, p. 163). A few years later, Goethe's disillusionment with garden art culminated in a clear favouring of the French garden style, because the architectural garden is clearly distinguished as a work of art lifted out from nature as non-art (cf. Buttlar 1989, p. 164).

In summary, it can be stated that in the course of the developments described, the constitutional scheme developed in art of a subjectively-aesthetically constructed visual space called 'landscape' in an image could also be extended to spatial contents of perception outside of art. Traditional visual expectations thus established a construction scheme that enables the viewer to *synthesise* the individual phenomena in a room or environment – as in landscape painting – into a pictorial visual whole (Burckhardt 1976 [1859]; Ritter 1996; Simmel 1990). Perceptually aesthetically constituted 'landscape' is thus a 'work of art in statu nascendi' (Simmel 1990; cf. Berr 2008; Piepmeier 1980) as the perception of a space as 'landscape' 'imitates' the construction scheme of art (Hard 1991). It is therefore a secondary derivation, because the word landscape is "secondarily projected onto reality, reified so to speak" (Hard 1991, p. 14; see also Hauck 2014; Kühne 2018f).

3.5   A German Affair I: The Influence of Romanticism

The construction of landscapes in the Romantic and Biedermeier periods, which were particularly intense and persistent in Germany, took a special path (Safranski 2007). In German Romanticism, influences from cultural, social, art, and ideological history flow together influencing the enrichment of the concept of landscape and everyday understanding through sedimented, multi-faceted connotations in a number of ways. The transition from agrarian to industrial society was associated with an alienation of man from his original attachment to nature (cf. Kühne 2018f, pp. 37–38). This is due in particular to technical and economic system logic: More and more people were no longer living in natural rhythms of the seasons (e.g. associated with sowing and harvesting) and the times of day (e.g. the presence of light for field work), but in rhythms that corresponded to the logic of economic and technical processes (e.g. shifts, production processes). Industrialisation was associated with social differentiation, which affected the contemporary understanding of the landscape by large sections of the population. The newly emerging world of work was characterised by increasingly complex, differentiated work processes and the emergence of correspondingly differentiated occupations, and this world remained "at least on Sunday, in leisure time, once again holistically experienceable in the form of the 'beautiful landscape'" (Bätzing 2000, p. 197). The former closeness to nature was replaced by a new scientific knowledge of nature and a stronger control over nature (see also Donnelly 2002). In this cultural and socio-historical context, the city becomes an obvious symbol of human rationality and a triumph of man over nature (Tuan 1979). From now on, landscape will be seen as a negative, emphasizing non-city and, in contrast to the city, symbolically connoted with freedom: "What drives the city dweller out of the door and into nature is precisely this: to escape the social constraints, the social and spatial narrowness of the city" (Kaufmann 2005, p. 59; see Tuan 1979). These developments ultimately led to a longing for nature, which in the eighteenth and nineteenth centuries was associated with a sentiment against cities of an urban educated middle class and a transfiguration of the rural (Kühne 2018f; Radkau 2002; Schenk 2013). In contrast to the city, landscape is stylised as rural-natural and in further developments as a "medium of social criticism" (Kühne 2018f, p. 47) particularly in 'Heimat' protection (see detailed Sect. 3.6), which will have to be returned to.

The aestheticization of the 'natural' as landscape – whether sublime as desert, sea, and high mountains or beautiful as cultivated by man in agricultural landscapes – also implies an emotional turn towards nature, which is also expressed in the enjoyment (Bourassa 1991; Howard 2011; e.g. Tuan 1979; Williams 1973; Haber 1993; Zapatka 1995; Cronon 1996; Fuhrer and Wölfing 1997; Fischer 2001; Hoeres 2004). The romantic aestheticisation and emotional occupation of sections of physical space that are seen together as a landscape can also be understood as a 're-enchantment' of nature that had previously been 'disenchanted' by the Enlightenment. Romanticism thus becomes the dark side of the Enlightenment (Illing 2006, p. 48; cf. also Sloterdijk 2007), for as a counter position it is constitutively related to the ideal of rationality of the Enlightenment. This demarcation from the Enlightenment led so far that Romantics conceived a holistic science of cognitive, moral and intuitive-aesthetic ideas (Eisel 2009; Eagleton 1994), an approach that is still used today in essentialist approaches to landscape (cf. Kühne et al. 2018) and in so-called 'holistic' scientific understandings of 'integrated research', for example in environmental research (cf. Gethmann 2005). The problem with holistic approaches is "that pretended insights of a holistic nature are particularly susceptible to ideology, i.e. that they are suitable for the practice of particularist politics under the veil of the surface grammar of a scientific speech" (Gethmann 2005, p. 392). In the nineteenth century, the *aesthetically* mediated construct of a holistically understood context of meaning 'landscape' was thus "transformed into an expression of good and true life in harmony with nature and the 'natural' social orders [and] under this anti-democratic perspective it was translated into a conservative political programme in the course of the Counter-Enlightenment and Romanticism" (Körner and Eisel 2006, p. 46).

In romantic landscape painting and literature, the revaluation of an emotional approach to 'landscape' plays a prominent role. In landscape painting, after the classical era, 'landscape' experiences "its highest revaluation in that mythological and historical contents merge into an expanded concept of 'landscape'" (Hohl 1977, p. 45; cf. also Piepmeier 1980). The picturesque representation of landscape was no longer merely a question of artistic practice, but "one of the inner and moral and religious constitution of the artist" (Büttner 2006, p. 262). In particular the works of Caspar David Friedrich (1774–1840) are characterised "so to speak by exuberant pathos" (Spanier 2006, p. 33) by dressing "spiritual and religious feelings in landscapes (landscapes of the soul)". In its allegorical and symbolic density, Romantic landscape painting with its frequent religious references (as in Friedrich's work) is partly tied to medieval depictions, but it is also an expression of the consciousness of the divorce of the modern individual from the context of paradisiacal nature (Zink 2006).

In romantic literature, landscape is "elevated to the status of a great symbol of the soul, as is the case with Heinse, Goethe, Jean Paul, Hölderlin, Eichendorff and Stifter. For Romanticism (…), the *subjectively unique atmosphere of* the landscape' is the basis of the perception of the landscape. Landscape representation is *'Stimmungskunst'* [English something like: 'sentiment art']" (Gruenter 1975a [1953], p. 207; emphasis: K.B. and O.K.) and thus tied to the emotionality of a viewer. Landscape as a mirror of subjective sensations is transferred as a 'Stimmungslandschaft' [English: 'mood landscape'] into the interior of the sensitive subject: "One of the most important chapters in the feeling of nature of the eighteenth century is the penetration of subjectivism into the depiction of the landscape, the 'Durchseeltwerden' [English something like: soul-seizing] of the landscape, the soulful relationship of man to the surrounding nature, that is, what one later tends to call a 'mood landscape'" (Langen 1975 [1953], p. 152). The prime example of this sentiment art is Goethe's *Werther* (Goethe 1985) as the work of *Sturm und Drang*: "The new soul-man in the new spiritual landscape, that is the basic attitude as in Werther" (Langen 1975 [1953], p. 160). Especially in the 'cult of sensitivity' (Doktor 1975; Krüger 1972; Sander 1974) this development leads to a longing for nature, which can hardly distinguish between one's own sensitive soul and the projection of mental states in a natural translation of a phenomena. The result is often a pantheism as a variant of holistic world understandings, i.e. an aspired natural-mythic fusion of 'subject and object', of 'I and world'in a 'Unio mystica' – nature and man seem equally 'inspired' to the romantic poets. Man is therefore "no longer separated from nature by anything, its plants and creatures like flowers and animals, a particle of the universe, blown through by its breath and animated by its soul" (Langen 1975 [1953], p. 170). The result was a cult of sensibility and feeling that was intended to ensure that nature and human feeling were in harmony. From this utopian content of an anticipated man-nature harmony a line of development can be drawn to ecologically inspired ambitions within the framework of an "aesthetic-political utopia" in the second half of the twentieth century and at the beginning of the twenty-first century, from the aesthetic "sign" of landscape to bring order and steer society "for the good of man and nature (…), at least to help steer it" (Hard 1991, p. 15).

In the Biedermeier era the interpretations and valuations experienced a specific twist: The restoration took place after the Congress of Vienna in 1815, and it was not until the revolutionary year of 1848 that the tensions between the bourgeoisie striving for co-determination and the aristocracy striving for absolutist rule eased. In the meantime, during the Biedermeier era, there were two options for people: Resistance or retreat into the private sphere. The followers of the Biedermeier era decided to retreat into the private sphere. The will to change the world via Romanticism transformed itself into a striving for a life in harmony and tranquillity within the immediate living environment. In the Biedermeier era, landscape became a symbol of humanity, which was seen as threatened in particular by the development of civilization and tendencies towards technical usability, general utility thinking, in physical space symbolised by the expansion of cities and industrial facilities (Kortländer 1977). In the Biedermeier era landscape was charged with the expectation of a kind of 'extended living room' in which peace and harmony were strived for.

Box 3: A legacy of German Romanticism: The recent discussion about 'atmospheres' in the landscape

In the semantics of the concept of landscape developed to date, a further line of development can be reconstructed, ranging from ancient, medieval, and modern Arcadian or paradisiacal associations, through literary 'moodscapes' (Langen 1975 [1953]) and Martin Heidegger and Friedrich Bollnow's concept of 'mood' to the currently inflationary concept of 'atmospheres'. Georg Wilhelm Friedrich Hegel (1770–1831), in his 'Critique of Romanticism' (cf. Pöggeler 1999), problematised a reference or even a reference to moods and feelings in the heyday of German Romanticism. There is the possibility that, in such a reference of sentiment, landscape can become a kind of atmospheric backdrop – in the sense of a 'mood landscape' (Langen 1975 [1953]) – and a landscape observer is motivated to an empty reverie, to a "reflection of his self without substance" (Hegel 1980, p. 12). A landscape execution induced by mood or sensation can thus lead to a landscape observer not going beyond the concentration on subjective inwardness. In this context, the term 'mood' can be seen as a 'psychological correlate' of the term 'landscape': "Landscape is thus, to put it simply, nature seen through a temperament, never nature itself, as an ontic" (Schneider 2009, p. 10). This assessment can also be found in the twentieth century with Martin Heidegger (1993 [1927]) and Friedrich Bollnow (1941).

A contrary consequence can consist in interpreting moods as moods of the landscape or nature itself, as prototypically in Carl Gustav Carus' (1789–1869) "Letters on Landscape Painting" (cf. Berr 2005, 2008; Carus 1982) is to be observed.

Hegel, on the other hand, deals with this question in relation to execution when he says that the 'soul' can be "intimate with natural objects when it is grasped according to some need", i.e. when they are felt" (Hegel 2003 [1823], p. 255). It can therefore only be spoken of a "character" (Hegel 2003 [1823]) of the landscape that corresponds to the states of the soul. Ultimately, it is a projection or construction.

It is this double or opposite problem of a 'subjectivism' or 'objectivism' of landscape or nature experience induced by a sensitive execution of landscape or nature, which is and remains virulent. This is evident in the current debate on 'atmospheres', in which the concept of 'atmosphere' appears as a counterpart to the concept of 'mood'. Corresponding attempts to determine the ambience in the landscape and to define its epistemological and aesthetic role can be found, for example, in Hermann Schmitz, Gernot and Hartmut Böhme, Michael Hauskeller and Michael Großheim. In essence, the debate revolves around the question of whether 'atmospheres' are projections or whether they belong to the landscape itself. The question is: "Does the landscape itself have the quality of being cheerful or melancholic, or do we attribute these qualities to it by projection?" (Großheim 1999, p. 325). According to Michael Großheim, representatives of such anthropocentric 'constructivism' or 'projectionism' include Georg Simmel, Ernst Bloch, Ruth and Dieter Groh, but also August Wilhelm Schlegel, Theodor Lipps, Martin Seel and Rolf Peter Sieferle. The main reproach to the so-called 'constructivists' and 'projectionists' is that they ignored the danger of a "complete loss of object" – that is, an object of cultural constructions or projections "over which cultural forms can first build themselves" (Großheim 1999, p. 359). The "historical research on aesthetics of nature", which investigates the "constitutional conditions of images, symbols, forms of appropriation of nature", would recognise the "subjective and objective spirit of images of nature", but would at the same time "eradicate every reference level and every content" (Böhme 1995, p. 139). Großheim, on the other hand, calls for a "phenomenology of atmospheres" (Großheim 1999, p. 361), which, in contrast to the theories emanating from the historically variable "constructedness" of the landscape, set a "history of fields of perception" (ibid., p. 357), as brought into play by Heinrich Rombach, for example (Rombach 1980). Constructedness is thus set against perception, as if perception can directly grasp atmospheres such as "objects of its own kind (…) that do not merge into projection" (Großheim 1999, p. 341), i.e. as "entities" (ibid., p. 343) or as "objective feelings" (ibid., p. 344), as they are ready for perception independently of perception. Recent debates on the concept of atmosphere, on the other hand, have endeavoured to convey the subjectivist and objectivist associations of this concept (for example Kazig 2007, 2013, 2019a; Hahn 2012).

3.6 A German Affair II: Landscape and 'Heimat' – The 'Heimatschutzbewegung' (English Approximately: Homeland Security Movement)

An examination of the typically German term 'Heimat' (Berr 2019a; Gebhard et al. 2007; Hinrichs 1974; e.g. Kühne 2018f; Kühne and Spellerberg 2010; Piechocki 2007, 2010, p. 152–162; Weber et al. 2019; Zöller 2015), which goes back for over a thousand years, unveils a broad spectrum of different meanings and reinterpretations. 'Heimat' etymologically derives from the Old High German 'Heim' having the meaning of 'settlement' and 'domicile', which has its linguistic origin in the Indo-European root *kei* with the meaning 'to lie' (Piechocki 2007, p. 20) – the word 'camp' ('Lager') therefore designates a 'place where one settles' (Kluge 1975, p. 299). The suffix formations *heimuoti* or *heimöti* always refer to a place of residence, a house or a property (Piechocki 2007, p. 20). As soon as a 'home' is built and basic human needs are satisfied, a "feeling of security and peace" arises (Piechocki 2010, p. 154). The "affiliation of persons and things" is addressed (Waldenfels 2005, p. 195) to a specific room or 'home'. Typical for this 'Heimat' as a feeling of belonging are terms like 'finding one's way home' (Hegel 2003 [1823], p. 105) or "to become at home" (Waldenfels 2005, p. 210) or "making oneself at home" in the sense of "subject-appropriate, *emotionally-practical* settling in" (Hinrichs 1974, p. 1038) into an environment. This feeling of security is made possible by the counter-experience of being foreign, uncomfortable, and threatened by others, strangers or outsiders after becoming settled and abandoning nomadic ways of life. From a *genetic* point of view, 'Heimat' therefore stands in conceptual contrast to 'foreigners' or 'distress' (Old High German: *elilenti*; Kluge 1975, p. 163) in the sense of 'other country' (Piechocki et al. 2007, p. 11). Living in a foreign country was therefore synonymous with living in distress. *Religiously* seen, concretely: in the Christianity of the early Medieval Ages, life on earth was *basically* a life in distress. The longing for home is then synonymous with the longing for the Kingdom of Heaven (Piltz 2007).

This religious meaning was gradually secularised from the twelfth century onwards, when the words *heimuoti* or *heimöti* were "related to home, wasteland, poverty and also family, trust" (Bertels 1997, p. 65). In the following centuries the semantic court developed this meaning into the sense of an objectively given "area of legal jurisdiction" (Greverus 1979, p. 64), with the concept of 'home' being replaced by 'ownership and entitlement rights (birthrights)' (Piechocki 2010, p. 154) in the context of medieval and early modern social orders in rural or urban communities. Until the middle of the nineteenth century, 'Heimat' was thus a legal concept and not yet a "category of experience of subjectively performed assignment to a socio-cultural space" (Greverus 1979, p. 64). There could be no question of subjectively perceived "sadness, poetry and sentimental splendour" (Jens 1985, p. 14). 'Home' as a legal term included rights and obligations, such as registration and deregistration obligations or the right to rules of procedure as well as provisions in emergency situations (Kühne 2018f, p. 272). With the increasing mobility of the emerging industrial society and the emergence of the German nation state, which henceforth assumed many of the tasks and obligations of the communities (Piechocki 2007, p. 24–25) the 'right of abode' as the "principle of stationary society" (Bausinger 1984, p. 13) lost its former function and significance. The concept of home could open semantically to new meanings.

This process of semantic reinterpretation took place mainly as a reaction to the negatively evaluated consequences of industrialization and the associated mobility of society, which was also associated with the loss of familiar domestic environments, in particular with the painting and poetry of Romanticism and in the middle of the nineteenth century with the emergence of 'Heimat' protection. With Romanticism, rural pre-modern forms of life, land use and housing were transfigured into the 'cosy and homelike' [*Heimelige*], 'Heimat' into the utopian "desired place of absolute security" (Hüppauf 2007, p. 116) or excessively idealised into the 'landscape of serenity' (Sieferle 1985). 'Heimat' as the original concept of law became an ideal idea of spatial belonging and thus *emotionalised* into a small-scale rural world, divided into small-scale social and political structures, unchallenged by the impositions of modernity (Kühne et al. 2018; Piechocki 2010; Schenk 2001b). 'Home' in this sense is the 'familiar world' (Waldenfels 2005, p. 198) which in Romanticism was increasingly emotionally merged with notions of 'landscape' (Weber et al. 2019).

In the eighteenth and nineteenth centuries, longings for an intact 'Heimat' and a premodern landscape merged with longings for nature (see Box 4), the transfiguration of the rural and an anti-city affection of urban educated citizens (Kühne 2018f; Radkau 2002) to an emotionalised context, on the basis of which developed a conservative 'Heimat' ideology and civilization critique (Körner and Eisel 2003). This 'Heimat' ideology, which was critical of civilization, leveled in connection with the idea of a unity of culture and nature as a typical regional unity of 'land and people' (Eisel 1980; Körner and Eisel 2006; Riehl 1854), i.e. as an "inextricable link between people and landscape" (Kühne 2018f, p. 40), paved the way for the 'Heimat' protection movement in the second half of the nineteenth century (Piechocki 2006). This figure of thought can be reconstructed as a fusion of connotations of the territorial ('regio') and aesthetic concept of landscape (Hard 1977; Schenk 2011). After Ernst Rudorff wrote his *Heimatschutz* (1994 [1897]) the idea quickly found popular acceptance and in 1904 the 'Bund Heimatschutz' (Association of Homeland Protection—protecting natural and historical areas) was founded (Piechocki 2010, p. 159). From then on, the rural landscape also became a protected object, currently within the framework of nature conservation, which is now also enshrined in law.

Box 4: The Germans and their forest

In Germany, forests experience a high social esteem. The (German) forest is attributed an identity-creating meaning (Lehmann 2001b, 2004, 2010). An essential element is the mystification of the Varus Battle in Romanticism: United Germans (or the Germanic tribes declared to be their ancestors), in combination with their forest, cannot be defeated even by the strongest army. Also rooted in Romanticism is the mythologisation of the (German) oak tree, which has been proclaimed "a symbol for eternity of the so-called Germanic people of origin" (Urmersbach 2009, p. 76). The words 'letter' and 'book' are based on the Beech tree, because of Germanic characters carved into beech sticks, and around 1500 place names in Germany also go back to the Beech tree. Even in the present day, forest is positively connotated with a 'natural state' (Jenal 2019; Jenal and Schönwald 2019; Kühne et al. 2014). On the other hand, forests are associated with people whose activities are considered socially deviant: The fairytale world with poachers, robbers, witches, fairies, and others (Urmersbach 2009) but whose activities are, in turn, associated with 'freedom' and 'independence'. In the course of the nineteenth century, the forest was attributed a political-pedagogical significance in the Romantic period and in the Biedermeier period as an individual norm, "may the forest improve man and his world" (Urmersbach 2009, p. 85). Accordingly, the migratory bird and 'Heimat' protection movements symbolically charged the forest as an alternative to industrialization, individualization, and economisation. The irony was that industrialization also took place through the substitution of fossil fuels for regenerative energy sources, which resulted in reforestation and forest conservation. What was idealised could be promoted because the things that were fought against were expanded (Schenk 2006; Uekötter 2007; Winiwarter and Knoll 2007; Zutz 2015). During the Nazi period, the myth of the 'German forest' was abused for propaganda purposes in the form of fantasies of superiority (cf. Körner 2006b). With the staging of the (German) forest in the sentimental regional 1950s Heimatfilm as a noble and wild piece of nature, should illustrated "the simple, so romantic, so beautiful, so completely apolitical" (Urmersbach 2009, p. 105) – beyond the Nazi's fantasies of omnipotence. In accordance with the high symbolic significance of the forest, the 'forest dieback' of the late 1970s and 1980s caused "a crisis of culture and influenced the current political consciousness of many contemporaries to a large extent" (Lehmann 1996, p. 145). Today the forest serves as an aesthetic landscape scenery, symbolically it stands for a harmonious coexistence, represented here by trees of different kinds and different ages (Lehmann 2001a). However, the high aesthetic esteem and emotional devotion are only hindered by a low level of cognitive knowledge (see also Jenal 2019; Kühne 2014a; Lehmann 2001b).

In contrast to the early conservative ideology of the 'Heimat' of feudal small states, which was captured in Romanticism, the 'Heimat' protection movement of the late nineteenth century was a reaction to the foundation of the 'Reich' in 1871 and its political-administrative centralism, which was to be countered by the compensatory diversity of regional 'land and people entities'. This essentialism of an understanding of landscape and 'Heimat' increasingly gave rise to folk terms such as 'Body of the People', 'Mother Earth', 'Fatherland' or 'Habitat' (Piechocki 2010, p. 155). After the Heimatschutz movement remained largely ineffective politically and in the period between the two World Wars the concept of Heimat was expanded from local to national (Kühne 2018f, p. 272) in the 1930s the National Socialist ideology adopted the folk-national and landscape connotations: "From the traditional ideas of 'Heimat', only the agrarian romantic aspect was adopted, while the regional component was strictly rejected. The national variant of home was emphasised" (Huber 1999, p. 47). The appropriation of the concept of 'Heimat' by a nationalistic and racist 'blood and soil ideology' has led to the political discrediting and even tabooing of the concept of 'Heimat' in Germany, which continues to this day. For example, in 1976 the term 'Heimat' was deleted from the Federal Nature Conservation Act (Bundesnaturschutzgesetz; BNatSchG 2009 [1976], § 17). Politically, the concept of 'Heimat' was finally appropriated by neo-Marxist tendencies, which followed on from Ernst Bloch – as a "cipher for a utopian state of fully successful unalienated existence, which for Bloch can only exist in a classless society" (Ott 2007, p. 60).

In the 1950s and 1960s, a quasi-depoliticised reanimation of the concept of 'Heimat' took place in the form of homeland films, 'Heimat' novels, and folk music via the newly created mass media. 'Home' as 'a collection of phrases and clichés, idylls and ideals' (Huber 1999, p. 48) was reduced to a kitschy, exclusively emotionally charged aspect. Since the 1970s, the emotional component of the concept of home has also been reactivated in another variant: In view of a diagnosed 'environmental crisis', 'Heimat' was reinterpreted as 'regional identity', 'homeland consciousness' as 'regional consciousness' (Greverus 1979). A superlative variant of the renewed emotionalisation of the concept of homeland is the 'Deep Ecology Movement', which was notably founded by Arne Naess (1973). Through a way of life which, being idealised normatively having to adapt to natural conditions, people can develop a new sense of home in the sense of a 'rooting' (Kühne 2018f, p. 273). 'Heimat' gradually was reinterpreted to the "active sphere of the ecologically oriented public" (Kühne 2018f, p. 272).

At the beginning of the twenty-first century, the references to earlier patterns of interpretation become more complex again. In particular, different politicisations can be observed: Citizens' initiatives, which were formed in the wake of the Energy Crisis, argue against the 'disfigurement' of their living environment strikingly often in similar noisy declarations in connection with 'landscape' and 'homeland' (Kühne and Weber 2019a; Weber 2018a). Moreover, it is political parties and associations that are reactivating 'homeland', on the one hand for localisation and on the other hand for the exclusion of others, which has become more or less explicitly intensified since the so-called 'refugee crisis' (Marg 2019; Reusswig 2019).

3.7 The Understanding of Landscape Today

Since the 1970s, Gerhard Hard, in particular, has conducted the first extensive studies on the non-scientific and pre-scientific understanding of landscapes and on colloquial and pre-theoretical associations of landscape perception and experience (Hard 1969, 1970c, 1977; Hard and Gliedner 1977). Hard was able to prove that with 'landscape' a "perceptual figure and conceptual figure" (Hard and Gliedner 1977, p. 18) is historically mediated and established, whose 'semantic court' (Hard 1969) is associated with 'landscape' that it is "natural", "rural", "all well and good", an "image", and a "stereotypical collection" of specific elements (Hard 1970c, pp. 25–97; see also Hokema 2013, p. 242 ff.; see Table 3.1).

Dorothea Hokema has conducted a comprehensive study (2013) of the more recent research results from other relevant studies used as a database and Hard's results as a "research matrix" for the question of whether the connotations that Hard had worked out as the colloquial core of lay people's understanding of the landscape "also influenced the understanding of the landscape at the beginning of the twenty-first century" (Hokema 2013, p. 242) (also Fig. 3.1). With regard to the different versions, uses, purposes, and contexts of use of the concept of landscape in German-speaking countries, Hokema distinguishes between a corresponding 'understanding of landscape' by 'laymen' and representatives of 'planning practice', by 'landscape concepts', and by 'spatial scientists'. The planning practice takes on a resource allocation and mediator position in so far as it, like the lay discourse, implicitly and explicitly takes up traditional concepts of landscape as well as how the spatial discourse seeks to transcend the pre-scientific understanding of life or to translate it into scientific terms. As a result of her study, Hokema was able to extract from the three discourses studied a 'core concept' of central concepts associated with landscape, shared by laypersons, planners, and scientists: "naturalness", "materiality", "normativity", "wholeness", and "social imprinting" (Hokema 2013, p. 279). This conceptual core is surrounded by a "conceptual environment whose contents can be consulted on a case-by-case basis" (ibid.). While now "the central concepts are directly or indirectly components of all landscape discourses examined and are therefore understood as necessary components of the landscape concept, the surrounding concepts are only used in certain discourses – they do not necessarily contribute to the definition of the landscape concept, but are compatible with it within the respective sub-discourses" (ibid.). Hokema points out that although the additional provisions in the graph can be loosely assigned from left to right to the discourse of spatial science, planning, or laypersons, no clear or fixed assignments of the additional provisions to the sub-discourses can be established. Also "clear demarcations between the sub-discourses are not possible" (Hokema 2013, p. 280).

The 'semantic court' of the concept of landscape that Hokema reconstructs in her study does not differ significantly from the results that Hard had already achieved in his

Table 3.1 Absolute and relative frequency of answers to the question "What other word do you think of first when you hear the word 'landscape'?" The survey compares the results of the 2004 and 2016 surveys and takes into account answers that were cited more than three times (Furthermore, adjectives in connection with nouns such as "natural environment", "free nature" etc. were not taken into account in the evaluation, only the nouns were taken into account. In the cases mentioned, "environment" and "nature"). The calculation basis for the relative figures is the number of citations above 3 in the respective reference year (according to: Kühne 2018e)

	Number of mentions 2004	Number of mentions 2016	Percentage of mentions 2004	Percentage of mentions 2016
Nature	100	121	27.0	34.8
Forest	59	38	15.9	10.9
'Heimat'	37	48	10.0	13.8
Meadow	25	18	6.8	5.2
Environment	20	11	5.4	3.2
Green	18	18	4.9	5.2
Mountains	18	10	4.9	2.9
Surrounding	18	13	4.9	3.7
Recreation	12	16	3.2	4.6
Fields	8	8	2.2	2.3
Neighborhood	7	5	1.9	1.4
Tranquillity	6	7	1.6	2.0
Hill	5	3	1.4	0.9
Trees	5	1	1.4	0.3
Vastness	4	1	1.1	0.3
Air	4	1	1.1	0.3
Beauty	4	7	1.1	2.0
Idyll	3	3	0.8	0.9
Tranquillity	3	0	0.8	0.0
Health	3	1	0.8	0.3
Garden/Gardens	3	1	0.8	0.3
Farming	3	5	0.8	1.4
Habitat	3	2	0.8	0.6
Running waters	0	3	0.0	0.9
Stagnant waters	1	4	0.3	1.1
Relaxation	1	3	0.3	0.9
Sum	**370**	**348**	**100.0**	**100.0**

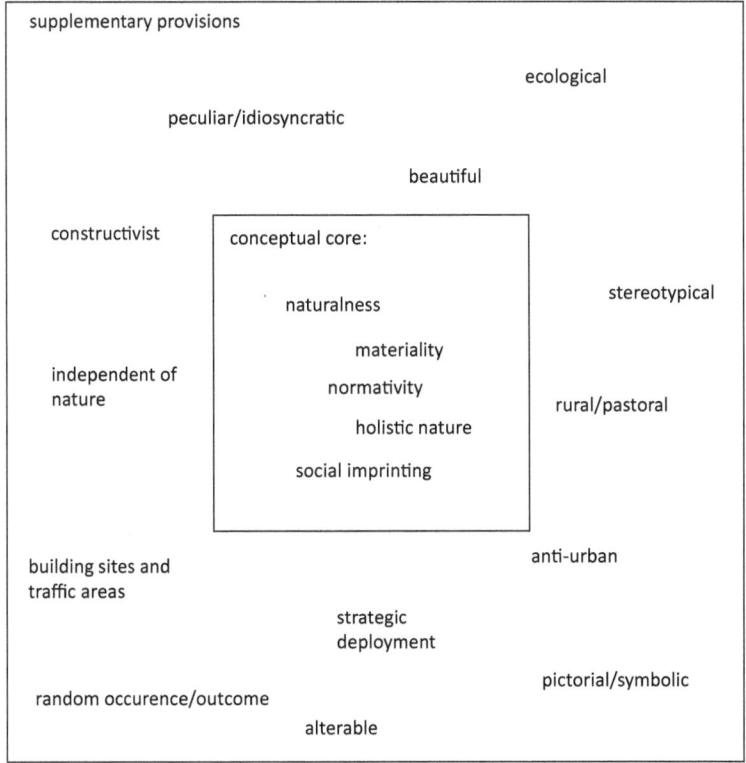

Fig. 3.1 The 'semantic core' and the 'supplementary provisions' of 'landscape' (graphically modified from Hokema 2013)

well-known studies. This is also not surprising, as even today the concept of landscape in the discourses and contexts of use examined is "put together from certain set pieces" by corresponding "individuals, disciplines, practitioners and other sub-discourses (…) [constructing] their understanding of landscape by using a pool of content determinations from which individual aspects are selected and put together like mosaics to form the respective concept of landscape" (Hokema 2013, p. 279). In this context, Olaf Kühne considers further semantic studies on aspects of the "social understanding of landscape" in addition to Hard's classical investigations and Hokema's current study (Kühne 2018f, p 47–53) and abstracts typical associations associated with the phenomenal object of landscape for many of the interviewees, which are reflected in the corresponding pre-scientific concepts of landscape. Kühne names and discusses the aspects of the "natural", "rural", "Heimat", "true good beauty", "figurativeness", "visuality" and "construction" based on a "stereotypical accumulation of elements" (Kühne 2018f, p. 48–50).

That "pool" as the "semantic core" of the concept of landscape seems, by the way, to be "extremely viscous" (Hokema 2013, p. 15). Because of the semantic sedimentation of a history of words and concepts documented since the year 830 (Gruenter 1975a [1953]), it comprises connotations, associations, and aspects that are still transported further in the various contexts of language, practice, and theory. These patterns of interpretation of landscape, which have been handed down and accumulated over 1200 years of cultural, social, intellectual, and art history, can also be seen in further targeted qualitative and, in particular, quantitative studies on the understanding of landscape in the German-speaking world – in the following using Saarland as an example. As a result of the two survey years (2004 and 2016), however, there are also changes (more on this in Kühne 2006, 2018e). To the open question of the household survey, "What other word do you think of first when you hear the word 'landscape'?" (Question 3), 408 of the 455 respondents in 2004 and 417 of the 450 in 2016 replied. 466 replies were given at the time of the earlier survey and 523 replies at the time of the later survey due to multiple responses. The most frequent answers (more than three mentions) are shown in the Table 1. In both reference years, the association most frequently mentioned is that of nature, with an upward trend. The association with forest/forests was found in second place in 2004, but in 2016, it was replaced and moved to third place by that of the 'Heimat'. The mentioned associations can be subdivided into more concrete terms (forest/forests, meadows/meadows, mountains/mountains) as well as more abstract terms (nature, environment[1]) and individual references to the world, e.g. as a recreational space, as a space of beauty and paradise as well as, in particular, of home. A clear shift in the references can be observed in the comparison of the two survey years: The proportion of abstract associations rose from 43.2 to 46.3%, while that of concrete associations fell from 38.9 to 29.3% and that of personal references rose from 17.8 to 24.4%. The association of landscape as a non-urban space is very clear in both reference years urban associations are largely absent.

What is striking about this study is the significant associative use of 'landscape' and 'nature', which was also designated and criticised by some authors as 'confusion' (Burckhardt 2006c; Küster 2009, 2012b). With a view to the social constructivist framing of the investigations in this book, such associations also lead not only to a dualistic 'concurrence' of 'natural' and 'cultural' connotations, but also to a reifying view of 'nature' as an unavailable foundation from which physical objects can be constructed into 'landscape'. 'Naturalness' is regarded by laymen, experts, and scientists as an "indispensable characteristic of the landscape" (Hokema 2013, p. 211) – whether as "wild nature" or as "tamed nature of the park or park-like landscapes" (Kühne 2013, p. 56).

---

[1]The clear dominance of the word "nature" over the word "environment" in the survey can be explained by the following factors Schemel et al. (2001) and Schemel (2004) from the clearly more positive resonance of the word "nature" in the form of "sensual and emotional associations" (Schemel 2004, p. 371), while more rational aspects are associated with the word "environment".

The "natural spatial features" are often "primarily attached to the 'green' in the sense of vegetation as well as to the colour 'green' as an indicator of the naturally grown" (Micheel 2012, p. 113). On the other hand, from the point of view of 'construction', even ecological landscape planning cannot reject the fact that "cultural landscape as an objectively describable physical section of the earth's surface" can at the same time be understood as "a socially constructed phenomenon of perception" (cf. Heiland 2010, p. 279). In perception, landscape is also "constructed as an object" (Kühne 2013, p. 57) and therefore 'conceived as designable and reparable', so that this view constitutes a 'condition for regional development (Micheel 2012, p. 113). Without such an understanding of the subject, "regional development measures could not, for example, relate to physical adaptation processes" (Kühne 2013, p. 57). The emphasis on this "double aspect" (Plessner 1981, p. 127) of 'Landscape' is legion. Landscape appears as "real structure" and as "objectified spirit" (following Martin Schwind, Hans Freyer, Nicolai Hartmann; Hard 2002b, vol. 1, p. 88); it "oscillates between a spatially-extended object and the image of a region" (Kaufmann 2005, p. 13). It shows an "irritating tension" because it is "at the same time an independent being and a mental construct" (Fischer 2007, p. 19). For the "culturally designed (…) can appear only on the ground of a given – nature as it is 'independent' of us" (Fischer 2004, p. 11). Thus, the landscape always contains "the physiognomic unity of natural conditions and their cultural forms as they present themselves to the viewer's gaze" (Sieferle 1997, p. 15). Accordingly, landscape stands "between physical-topographic factuality and cultural construction" (Seifert 2012, p. 19). In Cosgrove such dualisms culminate in the summarizing definition that landscape is "simultaneously a natural and a cultural space" (Cosgrove 2004, p. 68). However, the social constructivist separation of these two levels ('nature' and 'culture'), which was carried out at the beginning for methodological reasons, is clearly relativised in this way.

At this point, by the way, commonplace and scientific understandings of landscape already overlap. Against this background, the following chapter – after an excursion into the German-language understanding of common-sense landscapes in an intercultural comparison and the conclusion to this chapter – presents the development of landscape research in the German-speaking world – alongside classics of historical significance.

Box 5: The German-language understanding of Common-Sense Landscapes in an intercultural comparison

During the past decades, there have been several studies, some of them comparative (such as Drexler 2009b; Ueda 2009), to the linguistic and culturally bound understandings of what in German is referred to as landscape. The results are certain commonalities, but also clear differences, and even other modes of construction of substantive arrangements and symbolic charges. The aesthetic component is common with the English *landscape*, the French *paysage*, the Hungarian *taj*, but also with the Polish *krajobraz and* the German *landscape*. The substantive dimension contained in Hungarian and German (i.e. landscape as an object) is differentiated in English and French. In French, this component is covered by the words *pays*, *campagne*, and *terroir*. In French, the word *terroir* refers to the gustatory spatial location that is only marginally linked to the German concept of landscape. In English, the essential dimension is expressed through the terms *'land'* and *'country'*. However, these in turn form further 'semantic courts' (Hard 1969, 1970c), resulting in *'country'* as a political term. The German *landscape* and the Hungarian *taj* show a special emotional charge in the context of 'Heimat'. The Polish *krajobraz*, literally translated: View into the Country, is provided with a stronger subjective component than the German *Landschaft* (Drexler 2009a, b, 2013; Hernik and Dixon-Gough 2013; Olwig 2002). Even smaller are the intersections beyond the European languages: Arabic, for example, has no equivalent to the German *Landschaft*, but here the Arabic word for 'garden' is used provisionally (Makhzoumi 2002, 2015). In Turkish, on the other hand, the word *peyzaj* is used as a fiefdom word from the French language and follows on from the French meaning for the word *peyzaj* (Türer-Baskaya 2013). In Japanese and Chinese the references to material and symbolic spaces are on the one hand conceptually more differentiated, on the other hand the juxtaposition of nature and culture is not found, humans are regarded as part of nature, whereby a reference in the aesthetic mode of sublimity is also omitted (Taylor and Xu 2018). The concept of 'landscape' as a comprehensive aesthetic synopsis of the underlying material objects was first introduced as a scientific term through contact with Western science (especially German geography; Gehring and Kohsaka 2007; Taylor and Xu 2018; Ueda 2009, 2013; Zhang et al. 2013).

3.8 Bottom Line

Today's everyday speech about 'landscape' also draws on classical literature, historical art, plus scientific arguments, texts, and traditions – the historical and systematic connections of which should be known for landscape research. It can be seen how, for example, that certain notions or understandings of the term 'landscape' refer back to historically distant *topoi* and remain effective in their meaning to this day in a long traditional history of mediation. The following chapter shows that this tradition, which is often unknown or underestimated in its history, is also effective for scientific discourses and research in landscape-related sciences and disciplines. Some readers will miss one or the other classical text with regard to the everyday language as well as the scientific concept of landscape. A complete overview is naturally not possible in this venue. Moreover, such a complete overview in historical as well as systematic form is not yet available – there is an obvious desideratum of research here.

The Concept of Landscape in Landscape-Related Sciences

4

4.1   Introductory Remarks

If the development of landscape research in the German-speaking world is to be explained along the lines of its classics, then this combination of 'landscape research' and 'classics' needs to be explained. If it is possible to speak of 'classics' in a meaningful way, then in view of one of the "most important principles of science or scientific production" (Treibel 2004, p. 12), which generally applies to all sciences: Scientists "never start from scratch in their work, but always stand in a tradition – often unconsciously, because certain figures of argumentation have almost automatically entered their thinking" (Treibel 2004; cf. also fundamentally: Berger and Luckmann 1966). This also applies to landscape research, which is not only in the current conventional sense of research on 'landscape' as an object-level physical phenomenon or metastatic reflexive way of looking at or speaking (cf. Hofmeister and Mölders 2019; Kühne 2018d; Kühne et al. 2018; Kühne 2019c; Kühne and Weber 2019b; Kühne, Weber, Jenal 2019; Leibenath and Gailing 2012; Poerting and Marquardt 2019; Schenk 2017; Trepl 1996; Weber and Kühne 2019b), but also in the sense of the knowledge and consideration of texts and testimonies reaching far into the past, which thematise and describe spatial phenomena that we are prepared to grasp today under the term 'landscape', although such a subsequent attribution to these phenomena is often historically or factually inappropriate. Landscape researchers are also part of knowledge traditions that (co)determine their thinking and actions and should therefore be open to reflection. To this end, these traditions and classical figures of argumentation must, first of all, be known in their genesis, their history of impact and their claim to validity. In view of the immense variety of texts to which the title 'classical' could be assigned, a targeted selection must be made. The guidelines for this selection are decisive stages in the socio-cultural and, in particular, aesthetically well-founded history of the mediation of the term 'landscape' – as it

© Springer Fachmedien Wiesbaden GmbH, ein Teil von Springer Nature 2020   37
K. Berr and O. Kühne, *"Und das ungeheure Bild der Landschaft…"*, RaumFragen: Stadt – Region – Landschaft, https://doi.org/10.1007/978-3-658-27756-7_4

appears to us today to be taken for granted. First, classics of the thematisation of the discovery of the 'landscape' in modern times are presented. This is followed by references to classics on the conceptual history of 'landscape' as well as on the historicity of the socio-historically imparted physical landscape and the historicity of the execution rationality of this mediation – in the painting this execution rationality becomes apparent within the picture frame. The final part is made up of classical texts in the course of recent discussions.

4.2   Classics on the Discovery of the 'Landscape' and a Classic German Scientist Dispute

The fact that a physical-spatial section of the everyday environment could be seen *as a* landscape, at all, required specific prerequisites. Many researchers (e.g. Burckhardt (1976 [1859]; Stierle 1979; Jauß 1982; Steinmann 1995; Ritter 1974 [1963]; Schweda 2013; Vietta 1995) have cited the famous mountain ascent of Mont Ventoux by Francesco Petrarca on 26 April 1336 as an hour of discovery and classic topos of modern landscape observation in the sense of an impulse for a new aesthetic relationship with nature (Petrarca 1995). With Petrarcas mountain ascent the "in the Christian cultural circle" effectively historically decisive step for the discovery of a "view" on and over the landscape "as aesthetic value" (Wimmer 2004, p. 32) had been made. The classical study by Jacob Burckhardt on "The Culture of the Renaissance in Italy" (1976 [1859]) is exemplary for this point of view. Further studies following Burckhardt's interpretation are Joachim Ritter's now also classical essay on "Landscape" (1974 [1963]) and Karlheinz Stierle's study on "Petrarcas Landscapes" (1979; see also Blumenberg 1984; Dilthey 1914, p. 20; Gebser 1966; Jauß 1982, p. 140–142; Steinmann 1995, p. 39–49).

As a decisive characteristic of the new landscape consideration carried out for the first time by Petrarca (Ritter 1996), the "panoramic view" (Vietta 1995, p. 217) of a sentient and sensing observer (Ritter 1996) "into the distance, the actual landscape" (Burckhardt 1976 [1859], p. 274) is considered. This view into the distance and "of and across the landscape" was "decidedly no longer medieval" and "to be evaluated as an aesthetic cultural achievement of the first order" (cf. Vietta 1995, p. 216–217). A free and enjoyable observation of the landscape by a human being, as we take it for granted today in order "to be himself in nature" (Ritter 1974 [1963], p. 151), could not even be thought of in Petrarca's time. This new enjoying 'prospect' stood in contrast to the medieval prohibition of 'curiositas' (curiosity), here in the form of the 'concupiscentia oculorum' ('eye lust'; cf. Groh and Groh 1996, p. 29–31) rejected by Aurelius Augustinus: "Enjoy, *frui*, man may only God; everything else may only be used, *uti*" (Groh and Groh 1992, p. 294; cf. Flasch 1980). A summit ascent for sporting ambition or for the sake of an individual nature or landscape experience is still unthinkable in the ancient and medieval world. Mountain climbs that actually were carried out by the Greeks served mainly scientific purposes, by the Romans for political-military purposes (Steinmann 1995, p. 41), and by no means they were carried out for aesthetic reasons (cf. Gebser 1966). All the more astonishing are, therefore, Petrarca's remarks about his panoramic view from the summit of the mountain Mont Ventoux in the year 1336, which, despite the rather sparse depiction, let one feel something of the irritation of the new visual experience: "like a stunned man" he stands on the summit, "moved" by "the completely free panoramic view" (Petrarca 1995, p. 17) of the mountain landscape lying at his feet. At the same time, he is angry at himself because he admires earthly things for their own sake, instead of representing God in the sight of nature. Crushed to pieces, he returns to the traditional view of the soul and begins the descent. Petrarca's report thus remains "one of the great moments oscillating undecidedly between epochs" (Blumenberg 1984, p. 142). Petrarca stands on the epochal threshold between medieval and modern nature.

This new form of observation initially manifested itself in landscape painting (cf. Sect. 3.3), which only developed into an independent art discipline in the Renaissance (Burckhardt 1976 [1859]; Andrews 1989). According to Jacob Burckhardt, some "artists" already suspected and anticipated that "apart from research and knowledge […] there was another way of approaching nature […]. The Italians are the earliest among the modernists who perceived and enjoyed the shape of the landscape as something more or less beautiful" (Burckhardt 1976 [1859], p. 274). In Italy, a more lifelike representation of landscape developed, gradually replacing the medieval symbolic-allegorical representation, in which something depicted refers to something else, usually a religious meaning. Norbert Schneider (2009) has cited numerous examples of symbolically allegorical depictions in landscape painting. Examples are Giovanni Bellini: *Allegoria sacra*, c. 1480 (p. 30–36) for the quattrocento and Lorenzo Lotto: *Allegorie*, 1505 (p. 38–40) for the cinquecento.

However, the report and its standard interpretation are controversial. The critique concerns the dating of the ascent, the disregard of the difference between report and allegory, and the exact time of the epochal shift to the aesthetic experience of nature. Of particular note, it is debatable whether the epochal change had already begun in the fourteenth century or only during early Romanticism within the framework of modern physics theology. The reviews were presented, in particular, by Ruth and Dieter Groh (1992, 1996), Wilhelm Perpeet (1987, pp. 102–114), Martin Seel (1996, pp. 220-230) and Rolf Peter Sieferle (1986). Thomas Kirchhoff (2017), Mark Schweda (2013, pp. 180–194), Silvio Vietta (1995), Thomas Gil (2000) intervened as mediators in this dispute. The dispute can also be understood as a continuation of an older dispute, namely that between Heinrich Motz (1865) and Ludwig Friedländer (1867) in the sixties of the nineteenth century. This dispute concerned the aesthetic significance of mountain climbing in antiquity and the question of whether the ancient and medieval feelings of nature should be compared with those of modern times (cf. Falter 1999, p. 138).

The Groh spouses interpret the entire letter report of Petrarca as an allegorically disguised "conversion story" (Groh and Groh 1992, p. 293) in the tradition of Antonius and Augustine. The "image of the poet contemplating the world from above" is ultimately degraded "to a scientific mirage, a phantasm" (Groh and Groh 1992, p. 306).

The descriptions of the ascent, the mountain, the summit, the show, and the descent are all interpreted by the authors as metaphors and allegories in a medieval manner (Perpeet 1987, p. 111). Even the day of the ascent was wrongly declared, as it was postdated 17 years later (Perpeet 1987, p. 111). The report of the mountain ascent was indeed "rhetorically polished" (Perpeet 1987, p. 111), though "no letter" (Perpeet 1987, p. 112), but rather "a fictitious document of an idealised autobiography" (Groh and Groh 1992, p. 292; see also Groh and Groh 1996). The epochal newness of nature as a landscape experience is also denied. Nature is merely a metaphor for the earthly, and from a moral-philosophical perspective of the medieval motif of conversion from the exterior ("nature as a weak sign for everything that the soul of man wants to have, for all his obsessions") to the interior (soul), "nature becomes a fixed cipher in a moral-philosophical discourse" (Groh and Groh 1992, p. 301). An aesthetic experience of nature could therefore not appear at all in Petrarca's report. Perpeet also breaks with the standard interpretation when he simply denies Petrarca a "sense of free nature": for he had not discovered "nature as landscape" – "despite the contrary opinion of many of his interpreters" (Perpeet 1987, p. 104). On the contrary, landscape descriptions or a "feeling for scenic nature" (Perpeet 1987, p. 105) are equally missing in his poems. Additionally, the reaction to the alleged landscape experience described in the letter was rather poor and smelled "almost like paper" (Perpeet 1987, p. 113). Petrarca takes the spatial ascent to the mountaintop "in rhetorically trained gesture (…) to the entrance into the spiritual depth of self" (Perpeet 1987, p. 113). If at all, then Petrarca had "nature only in his back", one could by no means speak of "a love of Nature" (Perpeet 1987, p. 114). Petrarca's letter is therefore "to be doubted as a testimony of an epochal threshold with regard to landscape experience" (Groh and Groh 1996, p. 27).

There are also the proponents of the standard interpretation, like Stierle or Jauß, who have taken note of the newer Petrarca research and reacted to it. Ruth and Dieter Groh certainly concur this (Groh and Groh 1996, p. 27), but accuse the aforementioned of underestimating the difference between literary fiction and allegory, which ultimately "almost makes the 'real' mountain disappear in a network of metaphorical references" (Groh and Groh 1996, p. 28). Nevertheless, it must be asked whether the critics of standard theory do not go too far with all their true and justified criticism. There have been corresponding reactions to these criticisms, for example from S. Vietta and K. Steinmann. According to Vietta's counter-criticism, epochal upheavals always take place with many old and few new elements and figures. Decisive for this break is certainly the "panoramic view (…) of and over the landscape" (Vietta 1995, p. 217), which is indeed specifically modern and by no means medieval. Steinmann also defends the "spiritual-historical value of the letter" (Steinmann 1995, p. 48). Although "the medieval salvation care and world denial put the inquisitive curiosity of the Renaissance in its place", "the first step towards a new world view was taken" (Steinmann 1995, p. 49).

Another serious criticism concerns the exact time of the epochal change to the aesthetic experience of nature. In contrast to the standard interpretation, which identifies the beginning of this evolutionary history with Petrarca in the fourteenth century, Ruth and

Dieter Groh locate this beginning with the reinterpretation of nature within the framework of modern physics theology between the sixteenth and eighteenth centuries (1996). The formerly "terrible mountains", for example, were able to make nature appear as a "metaphor for the whole of a meaningful order" (Groh and Groh 1996, p. 112) in the physical-theological framework of the "figure of thought of the *oeconomia naturae*, of *design of nature*" (Groh and Groh 1996, p. 108; emphasis in the text) about its *use* (wood, sources, etc.) given to man by a Creator God, who expresses the greatness and wisdom of *God.* This had paved the way for an aesthetic experience of nature in the eighteenth century, which, however, also required the prior mediation of the reading of nature-describing texts and paintings depicting landscapes, which in the first place were able to establish specific views of nature as landscape. With this approach, however, there is a partial agreement between the advocates and critics of the standard interpretation, which can also be shown indirectly through an instructive statement by Thomas Gil: "The aesthetic experience of nature, which has been conceptually interpreted as an experience of 'landscape', is a very precondition-rich experience, which cannot always be had, i.e. not at any time. It requires a lot of things. It presupposes a series of subjective achievements of a mental nature. These subjective achievements are certain ideas and affects that are of an individual collective nature (…) The real experience of nature, of landscape is not a primary experience. It is preformed and mediated by ideas, conceptions, even worldviews or by the reception of works of art from poetry and painting" (Gil 2000, 56 f.). With the last two sentences, however, Gil almost literally paraphrases the conclusion of the Groh couple (1996, p. 114) at the end of their remarks on the transformation of aesthetic experience of nature in the output of physio-theological theories. This change in turn refers to a history of mediation as outlined in Chap. 3.

4.3 Classics on the Conceptual History and Historicity of 'Landscape'

In the German-speaking world are the works of Gerhard Hard (Hard 1969, 1970b, 1977; Hard and Gliedner 1977) on the 'semantic court' of 'Landschaft', the interdisciplinary anthologies by Wallthor and Quirin (1977) with the significant contributions by Gerhard Hard, Bernd Kortländer, and Gunter Müller, by A. Ritter (1975 [1953]) with important contributions by Rainer Gruenter and Ernst Cassirer as well as the Smuda anthology (1986). In the English-speaking world, the works of Appleton (1975, 1980), Cosgrove (1984, 1993), Jackson (1984), Olwig (1996), and Shepard (1967) are particularly noteworthy. As shown in Chap. 3, three main meanings of 'landscape' can be distinguished diachronically according to these preliminary works: Landscape as *regio*, landscape as 'picture' and 'soul symbol', and landscape as 'earth space'. Numerous landscape researchers have followed this preparatory work (cf. Haber 2006; Jessel 2005; Kirchhoff 2017; Kühne 2018 f; Leibenath and Gailing 2012; Schenk 2002, 2017 and many more).

For the most part, the *historicity of* the socio-historically mediated *physical* landscape was explicitly addressed by philosophers and sociologists in modern times. Johann Gottfried Herder was the first to explicitly address the historicity of culture and cultures in general (Herder 1964), as well as the historicity of the landscape in the sense of an essentialist (Kühne et al. 2018, 2019) unity of culture ('character of a people') and nature ('clima'; cf. Kirchhoff and Trepl 2009a, Kühne 2019c). In his in Germany well-known poem *Der Spaziergang* (English: the stoll; Schiller 2004a) and in his aesthetic letters (Schiller 2004b) against a sentimental unity with nature, Friedrich Schiller shows the sedimentation of cultural-historical processes in landscapes. Schiller's *idylls* therefore functions – in contrast to the idylls of Brockes, Geßner or Haller (cf. Sect. 3.3) – on the one hand as a counter image of a past harmony of man with nature to the present culture and society, as 'paradise', 'golden age', 'state of human innocence' *before* all culture. This is unreachable, because unrepeatable past is transferred from Schiller to the other as *télos*, i.e. as the last purpose of culture in a utopian future. Only the *hope* – "the idea (…) and the belief in the possible reality" (Schiller 2004c, p. 746–747) – of such a unique state of harmony of man with himself and with nature reconciles him with the unbearable civilizational and cultural evils of the present. In order to become effective, this ideological hope requires a "sensual affirmation" (Schiller 2004c, p. 746), i.e. an illustration, which, however, cannot be looked at in *real terms*, but can only be anticipated by means of representation in *poetry*. Schiller's idyll thus wants to lead man "theoretically backwards" by "practically leading him forward and ennobling him" (Schiller 2004c, p. 747). The poet should "lead us forward to our maturity", i.e. to a hoped-for perfection of humanity, he should "lead man, who can no longer return to *Arcadia*, to *Elysium*" (Schiller 2004c, p. 750), i.e. to a way of life oriented towards classical Greek poems. However, Schiller himself recognises that a return to antique models is not possible. What remains is the study of history in order to learn from it for the future (Schiller 2004d), as well as a shift in weight from content to *form* of reconciled relationships with regard to a "reasonable way of life" mediated by art (Gethmann-Siefert 1995, pp. 182–183).

Hegel discusses in his lectures on aesthetics (e.g. Hegel 2003 [1823]), in letters (Hoffmeister 1969, pp. 362–363) and other documents, cultural landscapes as an expression of human freedom and cultural history (cf. Berr 2005, 2009). Statements in Hegel's work that suggest little sympathy for the natural landscape can be found, for example, in the in Germany well-known *report on an Alpine hike* (1796) (Hegel 1989), in which Hegel talks about the landscape of the Bernese Oberland. At the sight of the rock masses, glaciers and waterfalls he can see neither something beautiful nor a feeling of sublimity, but the "sight of these eternally dead masses gave me nothing but the monotonous and, in the length, boring imagination: it is so" (ibid. 392). Otto Pöggeler therefore expressed the suspicion that "with [Caspar David] Friedrich searching for the misanthropic landscapes of the Giant Mountains, Hegel really didn't come to mind" (Pöggeler 2005, p. 237). Only the "free play" of moving water from flowing waters has "something charming" (Hegel 1989, p. 383), i.e. an aesthetic quality (Bondeli 1990, p. 213; Jaeschke

2003, p. 74). Martin Bondeli also pointed out that Hegel's *report on an Alpine hike* was "a counterpart, so to speak, to Albrecht von Haller's exuberant veneration of the Alps" (Bondeli 1990, p. 217–218; Jaeschke 2003, p. 75).

This early report with its predominantly "very prosaic" (Bondeli 1990, p. 214) descriptions of the Bernese natural landscape, on the other hand, is opposed by letters about a trip to the United Netherlands, which sound almost "poetic". During the outward journey Hegel enthuses about a place like Weilburg, which has "a romantic location", a "beautiful, narrow valley rich in vegetation, pleasant bends of the Lahn" (Hoffmeister 1969, p. 348–349). In the Netherlands he had "seen many beautiful cities, areas and paintings and churches"; Utrecht, for example, was "a beautiful city with a university and graceful surroundings" (Hoffmeister 1969, p. 363). On his way back he crosses German heath landscapes and complains wistfully with an "goodbye beautiful Holland and Brabant, from here all pagans"; and then unfortunately "day and night went through barren heaths" (Hoffmeister 1969, p. 365). If the "surroundings of Osnabrück" are still "quite graceful", Hegel regrets having to drive over a path with Diepholz in the most beautiful sunshine, on which the sun must "shine on such steppes". "But against Bremen" he then discovers, almost relieved, "Dutch green meadows" again (Hoffmeister 1969, p. 366), as he notes at the end of the letter to his wife. The Dutch urban culture with its "graceful surroundings", its "green meadows", its "beautiful areas" is therefore for Hegel the epitome of a culturally mediated "liberation" and treatment of nature, in which cities are embedded in cultural landscapes. At no point does he attribute "beauty, not even sublimity" to nature (Jaeschke 2003, p. 75). Such attributions can only be found in the statements about cultural landscapes, which he believes to be found in Germany, but above all in the Netherlands. Of course, Hegel overlooks the fact that the "barren heath" is also a landscape in so far as it is the product of human forms of agricultural cultivation.

As in Hegel's work, Theodor W. Adorno also conveys nature through historical processes (Adorno 1970a, pp. 97–121), and landscape is also shaped by this historical mediation, since "history as its expression, historical continuity as form, has become engraved in it" (ibid., pp. 101). For the category of natural beauty this means: "Natural beauty is suspended history, pausing becoming" (ÄT, 111). Regarding the aesthetically relevant relationship between art and nature, Adorno claims that these are "referred to each other: Nature on the experience of a mediated, objectified world, the work of art on nature, the mediated governor of immediacy" (Adorno 1970a, p. 98). The present total mediation between nature and beauty of nature, however, leads to the fact that beauty of nature "passes over into its grimace" (Adorno 1970a, p. 106) – Adorno brings as an example the supposedly beautiful girl's face, which, however, merely imitates the beauty standards of a film star conveyed by the film industry. The 'reality', which Hegel understood as historically mediated and reasonable reality (cf. Hegel 1995), is regarded by Adorno as 'false totality', as 'delusional context', which had historically arisen through the compulsion of the 'identifying thinking' of 'instrumental reason' (Adorno 1970b; Birnbacher 1992). Adorno sees as the last possibility, to hold up a mirror of reality's alienation from reality, as the only the recourse to natural beauty and the imitation of nature rejected by Hegel. The beauty of nature is a beauty that man has not produced and is not dominated by man. Only in the aesthetic experience of nature or natural beauty – for example in the form of a 'beautiful landscape' – can the continuum of social rule be broken through. The imitation of the beauty of nature is not a copying reproduction of natural phenomena, but an imitation of something unrestrained, unproduced, of something that is not integrated into the 'blinding context' of the 'false totality' – it is "imitation of an inimitable" (Gethmann-Siefert 1995, p. 237).

The historical nature of the *execution rationality* of socio-historically mediated physical landscapes was also addressed at an early stage. Three classics should be mentioned as examples. Immanuel Kant as the 'forefather' of constitutional landscape theories was able to show that and how the perception of landscape phenomena as 'sublime' presupposes a distance from 'nature' and therefore a cultivated, historically developed reason (Kant 1993, p. 23–29). Kant adopts the category of the 'sublime' from the empirical tradition and gives it a new meaning, in that nature, regardless of its sensually incomprehensible greatness (mathematically sublime) and its horror (dynamically sublime), opens up to man an experience of his freedom in contrast to being at the mercy of nature. With regard to the question of whether the experience of the sublime is equally possible for all human beings, Kant assumes on the one hand that only such a culture, which has already progressed far in the cultivation of reason, can at all give a central meaning to the feeling of the sublime: "Indeed, without the development of moral ideas, that which we, prepared by culture, call sublime will appear merely deterrent to the raw human being without the development of moral ideas" (Kant 1993, p. 29). On the other hand, the experience of the sublime is not simply "convention", i.e. contingent agreement in a culture and society, but all people in every culture possess in principle the potential of reason and thus the experience of the sublime. "But for this reason, because the judgement of the sublime requires culture from nature (…), it is not, after all, produced by culture first and merely conventionally introduced into society; but it has its basis in human nature, namely that which one can simultaneously make sense of and demand of everyone with a sound understanding, namely in the attachment to the feeling for (practical) ideas, i.e. to the moral" (Kant 1993, p. 29). Kant thus builds a bridge between the aesthetic and the moral. In colloquial language, but also in scientific discourses, the 'beautiful' and the 'good' are often used, which leads to an amalgamation of aestheticisation and moralisation tendencies (cf. Berr 2019c; Berr and Kühne 2019a, 2019b; Kühne 2018c, 2019g; Weber et al. 2018) and thus leads to a complexisation and complication of landscape conflicts (cf. Berr and Jenal 2019; Kühne 2018a; Weber 2018a; Weber, Jenal et al. 2016; Weber and Kühne 2016). Finally, the two sociologists Hans Freyer and Lucius Burckhardt are to be mentioned, who discussed the historicity of the execution relationality of the 'reflexive entity' (Freyer 1990) or the 'transitoriness' of the landscape (Burckhardt 2006c).

This execution rationality of 'landscape' is shown in painting as an explicit *setting*. The artist must create an excerpt from the world, the picture has to limit the field of vision. The device for this is the picture *frame* (cf. Hegel 2003 [1823], p. 44). The "accentuation of the Execution Character" (Gethmann-Siefert 2005, p. 281) of the perception of landscape is associated with this. The frame is necessary because painting reduces the three-dimensional space of architecture and sculpture to the two-dimensional surface. This is associated with the fact that the interplay of light and shade is the "curvature of the figure" (Hegel 2003 [1823], p. 250) by the means of painting itself. A three-dimensional, palpable shape becomes "superfluous" (Hegel 2003 [1823], p. 251). Light and dark must therefore not be left to chance – as is the case in nature – but painters must "fix the light and dark", whereby a "background is necessary" (Hegel 2003 [1823], p. 251). The foreground, middle ground and background are created. The framed surface of the picture is thus able to unite the figure (sculpture) and the background (architecture) on one level. As far as the colour is concerned, it is only in the 'successive execution' of seeing that it brings the forms, the sculptural shapes, and the spatiality "through imperceptible transitions of one colour into the other" (Hegel 2003 [1823], p. 261) into being. On the part of the subject – who sets the picture frame and thus a conscious artistic design – this corresponds to a spiritual sensuality that frees the subject from all haptic, i.e. still somehow connected to the sense of touch. For what appears on the picture surface as the representation of a content is constituted only in the 'execution of seeing'. Through the colour and the frame, the form is constituted for the eye; the 'construction' of a world in the picture takes place. In this way the 'conscious design' of a subjective, but socio-culturally mediated world view is created. Hegel's aesthetics of content, which integrate reception and production aesthetics, starts from the artist's ability "to convey (production), through an *individual* experience of the world, something 'all-too-human' (possibilities) that every human being has in store for *us* to understand (reception)" (cf. Berr 2013, p. 160).

The concept of the draft refers to the concept of possibility (cf. e.g. Berr 2013; Hubig 2013; Poser 2004), i.e. that something is highlighted (framed, cut out) from a given space. In a lecture from 1937, the art historian Kurt Bauch (1957 [1937]) pointed out the outstanding importance of seventeenth century Dutch painting for the image of the world and the image we make of ourselves in modern times. With a view to Dutch landscape painting of the seventeenth century, Bauch shows that these pictures – as well as the genre and still life pictures – are concerned with the design of possibilities for worldview and interpretation: "Now, however, a portrait of nature is given, no longer reduced in size qualitatively, but quantitatively; a part of it, a section is brought (…) Only the individual can be grasped, it becomes accessible as part, as section. Our contribution is not what the ancients or we think, but only what we cut out, where we turn, what we 'consider'" (cf. Bauch 1957 [1937], p. 127). In this way, *new* perspectives, views, perceptions, overviews, new objects and new themes emerge. Images cut out certain "views" from the context of the given, they open up access to "possibilities of reality" (Grassi 1962, p. 36). They offer designs of the world and reality that reflect man's *historical* relationship to his environment. Such images open up a space of the possible within reality. These possibilities are partial aspects of reality, which in this way enter the consciousness of the image producers and recipients through the image.

Like Kurt Bauch, Georg Simmel also left a classic text on the function of the 'picture frame' (Simmel 1995 [1902]). Wilhelm Heinrich Riehl (1996) has shown that it is necessary to train the described compositional mechanism in order to examine the landscape and that the well-known formulation of a 'landscape eye' has been coined for this purpose. Simmel described this compositional contribution as a synthesis in the sense of a synopsis of "individual objects of nature" (1957 [1913], p. 147), which makes it possible to view the product of this synopsis "as landscape" (Simmel 1957 [1913], p. 142).

4.4   Recent Discussions

If something currently unites the landscape-related disciplines, then it is the disagreement over the meaning and scope of the object landscape, which, as has been shown, has already been interpreted colloquially, as well as scientifically, into an almost incalculable variety of determinations, aspects, and interpretative approaches. "Landscape" connotes such an abundance of characteristics that almost everything is addressed or the "flag word 'landscape' (...) means everything and nothing" (Hard 1977, p. 22). 'Landscape' therefore represents one of the "extremely unclear concepts of the European political and intellectual history of the last millennium" (Hauser and Kamleithner 2006, p. 74). This "meaningful vagueness" is then also the reason why "the concept of landscape takes up both classical tradition and facilitate innovation" (Hard 1977, p. 15). "Flag words" (cf. Dieckmann 1975) are mostly positively connoted and normatively charged (cf. Klein 2005, pp. 128–140) and can therefore – depending on the interest in knowledge and research plus depending on theoretical, methodological or methodical background – be used for whatever interests and purposes and potentially instrumentalised ideologically or power-oriented. Diedrich Bruns and Olaf Kühne (2013) tie in with Detlef Ipsen, for whom this term is a 'compositional' one, in the view of the historically conveyed richness of meaning of the concept of landscape (Ipsen et al. 2003, p. 13). It contains "life-worldly, aesthetic, territorial, social, political, economic, geographical, planning, ethnological, but also philosophical references, and with each of these references its content can be changed" (Bruns and Kühne 2013, p. 15). Winfried Schenk therefore also speaks of the "vagueness of the term" landscape (Schenk 2002, p. 7). If one summarises with Schenk the linguistic-historical and cultural-historical research, "landscape" is a "tinted" term "with a multitude of connotations, with which it eludes a generally valid definition" (p. 7), "due to its holistic character (...) it cannot be operationalised" (p. 12), and "thus [seems to be]completely unsuitable for use in science" (p. 7).

However, while Schenk sticks to the concept of landscape as a term suitable for scientific purposes, Ludwig Trepl (1996, 2009), Gerhard Hard (1970c, 2002a), and Karlheinz Paffen (1973a), for example, expressly doubt this. Thus, with a view to discussions in landscape architecture, Trepl expressly opposes the attempt by Martin Prominski (2004) to take over J. B. Jackson's (1984) concept 'Landscape Three' so to define a "term for the totality of landscape architecture design objects" (Trepl 2009, p. 327). Not only did this attempt fail, but it seems to Trepl "very questionable" whether this "brings much" at all (Trepl 2009, p. 327). If such a concept is to be developed and defined, then "a concept for landscape that can be used for landscape architecture must be *developed from* the everyday language, if that should be necessary" (Trepl 2009, p. 327). In view of the ambiguity of the term landscape, Hard proposed that it be used only 'colloquially for' geography (Hard 1970c). At the same time, he turns against the determination of essence, but also against a phenomenology of the landscape (Hard 2002a), since in his opinion these merely amount to "language-induced evidence experiences of the highest triviality" (Hard 2002a, p. 84). Paffen (1973b) rejected this term roundly as scientifically inexact. By charging the pre-scientific as well as scientific use of language with predominantly positive associations ('beautiful landscape') and by the range of the 'semantic court' (e.g. Hard 1970c, 1977) of the landscape term and thus manifold possibilities of identification, the term 'landscape' can thus be for some "a grateful term with a high sympathy value and low risk of definition" (Franzen and Krebs 2005, p. 25), for others it is doubtful whether it is suitable for scientific purposes.

Further texts of the *more recent* discussions about 'landscape', which can certainly be classified as 'classical', deal on the one hand with the supposed 'end of the aesthetic category landscape' (Piepmeier 1980) – either in the sense of dissolution into a 'Landscape Three' (for the English-speaking world: Jackson 1984; for the German-speaking world: Prominski 2004), in a 'total landscape' that has been completely humanly transformed and can no longer be planned (Sieferle 1997), or in a hybrid spatial structure that overcomes the traditional difference between city and countryside, which Thomas Sieverts called 'Zwischenstadt' (English approximately: 'Intermediary-City'; Sieverts 1998 [1997]). At the latest since the publication of Thomas Sieverts' book on the 'Zwischenstadt', voices from various disciplines have increased that classify the urban-rural difference as historically outdated, empirically cancelled, theoretically useless or practically irrelevant. Sieverts characterises this type of settlement as a city "between the old historical city centres and the open landscape, between the place as living space and the 'non-places' of overcoming space, between the small local economic cycles and dependence on the world market" (Sieverts 1997, p. 7). In a nutshell, the following diagnosis is always presented in numerous variants following Sieverts: In the history of the urban-rural relationship there was an urban-rural contrast, which on the one hand was characterised by the 'city of containers' or 'European city', and on the other by the 'agricultural landscape' (Sieferle 1997) or traditional cultural landscape. This relationship has changed since the Industrial Revolution, and since the beginning of the second half of the twentieth century, the old urban-rural contrast seems to have simply 'disappeared in real' terms or to be disappearing in the intermediate city structures.

Many authors now claim that the cultural developments described in recent decades have inevitably led to the loss of landscape as a reference space for nature (Schneider et al. 1992, p. 177). Countering Joachim Ritter (cf. Sect. 4.3) Piepmeier (1980) stated that the aesthetic concept of landscape, based on the separation and juxtaposition of city and appropriated nature on the one hand and 'free nature' not appropriated by man on the other hand, loses its foundation and its justification when nature is completely appropriated by society. But this is the empirical finding in Central Europe since the Second World War: There is no longer any nature untouched by human intervention, and thus no non-cultural counterpart to culture and society that can aesthetically convey nature to man. At the end of the 1990s Werner Bätzing asked a somewhat different question, namely whether rural areas are simply 'disappearing' in view of the dramatic changes since industrialisation, and especially after the Second World War. Bätzing nevertheless speaks of what he calls 'urban-rural complementarity', the final dissolution of which could lead to an 'inhospitability' (Mitscherlich 1965) of both cities and rural areas (Bätzing 2002), in view of the empirically established tendencies towards dissolution of the urban-rural contrast. However, the majority of the authors do not continue to speak of 'urban-rural contrast' or 'urban-rural relationship', but in the current discussion terms such as 'patchwork landscape', 'net city', 'suburban space', 'urban landscape', 'periphery', 'settlement mash', etc. are used (cf. Vicenzotti 2011, p. 74; Vicenzotti 2019), *Stadtlandschaft* (English: Citylandscape; Hofmeister and Mölders 2019) or 'Stadtlandhybride' (English: urban-rural-hybrids), Spatial Pastiches and URFSURBS (Urbanizing Former Suburbs; cf. Kühne and Weber 2019b).

The urban-rural relationship that has become questionable and can be problematised can now evoke various antitheses, i.e. be interpreted and evaluated differently, depending on what prior understanding or basic assumptions the respective scientists involved in the planning and design process or who are concerned with this problem associate with 'city' or 'landscape'. Accordingly, there are theorists who reject the phenomenon of the 'Zwischenstadt', those who welcome it, and those who want to accept it and actively shape or 'qualify' it. Vera Vicenzotti has meticulously examined this *Zwischenstadtdiskurs* (Intermediary-city discourse) in connection with preliminary work by Gabriele Schultheiss and has summarised these three possible positions under the ideologically typical generic terms 'opponent', 'euphoric' and 'qualifier' (cf. Vicenzotti 2011, 2012; Schultheiß 2007). In contrast to the traditional urban and landscape concepts, the 'opponents' associate the *Zwischenstadt* with a lack of history, identity, and therefore rootlessness, as well as a fragmentation and heterogeneity of the residential space. They speak out against mixing, but in favour of structural, social, and cultural density. The 'euphorics' speak out against linking history and local identity and instead focus on indeterminacy and openness, fragmentation and heterogeneity. They rely on the design of the fragments instead of the wholeness. They are in favor of unlimited possibilities and mixtures of possible social interests. The 'qualifiers' speak of a 'valorisation' or 'qualification' of the intermediate city structures, wanting to shape them into urban regions full of character, for example with historical elements as 'identity anchors'. The spatial breaks can either be 'reconciled' or 'staged'. In view of an accepted plurality of social interests, it is a matter of their placement in space as well.

The three attitudes are each subject to specific criticism (cf. Schultheiß 2007; Vicenzotti 2012): In view of the actual settlement events, the 'opponents' are accused of an anachronistic and ideologically susceptible position and of overestimating the planning possibilities. The 'euphorics' are blamed for the fact that design is impossible with them, since indeterminacy cannot be created. They are also accused of having a 'neo-liberal view' because they are indifferent to social concerns. In terms of regional planning, they thus violated the German Spatial Planning Act (Raumordnungsgesetz; ROG), which requires equal living conditions in all rooms. The fundamental dilemma of the 'qualifiers' is the "contradiction between the alleged acceptance of the aesthetically perceived spatial matrix" – i.e. the intermediate city structure – and the "aesthetically defined concept of order of a 'European cityscape'" (Schultheiß 2007, p. 102). The "questionability" and the frequently observed failure of the qualification strategy are "related to the fact that it wants to put its claim to design into practice" (ibid.).

This discussion is symptomatic of a frequent confusion between an object-level reifying view of 'landscape' and a meta-level reflection on the socially and culturally mediated meaning of 'landscape' (cf. Körner 2006a). The basic concepts of essentialism, positivism, and constructivism of the theory of science in the field of tension (cf. Kühne et al. 2018) oscillates between reification and reflection on meaning (cf. Kühne 2019c).

On the other hand, more recent discussions revolve around the connection between 'landscape' and 'Heimat' (cf. Hülz et. al. 2019; Kühne 2009a; Kühne and Spellerberg 2010) or 'Heimat' and 'nature conservation' (cf. Körner et al. 2003; Körner 2006b; Weber, Kühne, Hülz 2019; cf. also Hupke 2015, 2019a, 2019b, 2019c; Körner 2005a), which in particular have to deal with the traditional understanding of 'Heimat landscape' as a counter-concept critical of civilisation. Decisive classics of this tradition are the already mentioned treatise by Herder (1964) as well as the historically significant writings by Hugo Conwentz (1904), Ludwig Klages (2013 [1913]), Wilhelm Heinrich Riehl (1854), Ernst Rudorff (1994 [1897]) and Paul Schultze-Naumburg (1901–1917).

4.5 A German Special Path: Landscape in German Geography, Illustrated by Comparison with Anglo-Saxon Geography

The development of the 'expert special knowledge stocks' on the subject of landscape is both subject-specific and differentiated, but there are also culture-specific influences (through certain landscape concepts), national research traditions, and influences of certain research personalities. In addition, there are specific discursive developments as well as scientific and educational policy framework conditions (such as the priority given to certain disciplines or whether and with what intensity they are taught in schools). In the following, the path of Anglo-Saxon geography in comparison to German-speaking (especially human) geography will be outlined with the aim of contextualizing certain developments and logics of German-language landscape research. An essential characteristic of German-language human geography is that it has developed largely self-referentially for a very long time (up to the 1990s), without attaching particular importance to interdisciplinary networking or conceptual input from foreign-language geographies (Gebhardt 2019; Hard 2002c; Schultz 1980; Wardenga 1989, 2001). Accordingly, there are lines of development and ruptures here that were not, or were differently, carried out in Anglo-Saxon landscape-related human geography. In this respect, the developments in Anglo-Saxon and German geographical landscape research over the past 100 years are presented in the following exemplary manner. This comparison acquires a special charm because, at the beginning of the scientific examination of landscape, research by German-speaking researchers was of great importance. A relationship that has reversed today. The Anglo-Saxon geography is differentiated into a British and a North American tradition only to a limited extent, since the focus of the comparison lies on German-language human geography landscape research (more detailed descriptions of the respective traditions can be found in Winchester, Kong and Dunn 2003, Wylie 2007, Duncan and Duncan 2009; Hard 1970c, 1977; Schenk 2013; Wardenga 2006; in a summary for Antrop 2018, a less detailed comparison than the following can already be found in Kühne 2019c).

The roots of Anglo-Saxon (especially North American) and German landscape research can be found in the understanding of landscape by Carl O. Sauer (1889–1975), a German geographer. He built the Berkeley School of Geography into one of the leading institutes between the 1920s and 1950s. His central research object was the (cultural) landscape, whereby he internationalised the concept of landscape of the 'classical' German geography from the turn of the nineteenth to the twentieth century (Price and Lewis 1993). In his classical definition Carl O. Sauer (1969 [1925], p. 46) writes: "The cultural landscape is fashioned from a natural landscape by a cultural group. Culture is the agent, the natural area is the medium, the cultural landscape is the result" (see detailed: Mathewson 2009). Here the close connection between culture and nature becomes clear in the formation of landscape, which he understood as a real object, which also points to positivistic components of his understanding of landscape. However, he also understood 'cultural landscape' as a 'superorganism' in which culture had a strong influence on physical space, which in turn induced essentialist components of his understanding of landscape. With his understanding of the landscape, he set himself apart from the previously dominant geo-determinism (which assumed that natural factors determined culture), with which he found himself in a long tradition of holistic science in Germany: "Sauer's view of science can be traced in large part to his devotion to Goethe's humanism and the historicism of German geography" (Duncan and Duncan 2009, p. 2). The classical geographical concept of landscape at that time arose from two traditions of thought (Hard 1977, p. 15) "1): the 'physiognomic' tradition of the versatile interested traveller combined from 'naive' world view and 'scenic eye' and 2) the 'regionalistic' tradition of 'thinking in earth spaces' and earth space divisions". This quote illustrates the essentialist-positivist hybridity of the research tradition, which was not limited to human geography, but also met other 'sciences on the move', such as sociology (for example in the person of Vilfredo Pareto; for more see Albert 2005). Oppel, as a typical representative of this tradition, could define landscape as an "earth space which presents itself as a whole from any point of view" (1884, p. 36), Alexander von Humboldt was able to assume an access to the "character" of a space beyond the visual with the ascribed sentence of the "total character of an earth region" (Hard 1970b). This expression of the 'character' points to an essentialist understanding of von Humboldt as well as the holistic naming of a 'totality', because the access is not positivist-dividing, but takes place 'holistically'. Thirdly, the term 'earth region' refers to the presence of other 'earth regions', which are correspondingly endowed with a different 'total character' (Kühne 2019c).

Similar to the 'classical' German cultural geography, the members of the Berkeley School and those who affirmatively referred to it were also strongly oriented towards the investigation of culturally conditioned inscriptions into physical space understood as landscape (Mathewson 2009; Wagner and Mikesell 1962). According to the essentialist core of this view, the focus was on the investigation of traditional material elements, such as the mapping of house, roof, and corridor forms (e.g. Born 1977; Kniffen 1965), whereby the basis of the typification underlying the mappings is not to be found in an individual phenomenological experience, but in abstractions based on empirical and historical investigations (Duncan and Duncan 2009; Gebhardt 2016). Thereby 'landscape' (understood as a physical object) was increasingly framed with the metaphor of the "'palimpsest' (documents partially erased and overlain with newer forms and patterns) holding a wealth of information and clues to their histories by those who were able to recognise significant features and relate these to a larger system of landscape features" (Duncan and Duncan 2009, p. 4). In Germany, for example, Hartke (1956) analysed the material-landscape succession on agricultural land as a 'Sozialbrache' (English approximately: 'social wasteland') as a result of socio-economic processes of the growth of structural change in rural areas, i.e. ultimately as inscription. The spatial image of 'traditional geography' was a "well-ordered mosaic of spatially segmented natural and social units" (Blotevogel 1996, p. 13) which in turn refers to essentialist patterns of interpretation. The access to the physical spaces interpreted as 'landscape' was characterised by 'expert' special knowledge stocks. Wylie (2007, p. 41) characterised the expert understanding of Sauer and his successors as follows: An "expert, someone who stands apart from the phenomena in question, the better to objectively scrutinise it".

The criticism of the 'classical landscape paradigm' was comparable in content both in the Anglo-Saxon region and in Germany. However, the consequences for the further development of the subject differed fundamentally. The classical paradigm was criticised:

1. the holistic organizational understanding of landscape and culture, which, for example, left no room for individual development (e.g. Duncan 1980),
2. the 'object fetishism' (Duncan 1990, p. 11), i.e. the focus on material aspects of culture,
3. the focus on research into 'traditional' and 'exotic' landscapes (Winchester et al. 2003) while largely ignoring urban developments (Mathewson 2009),

4. the structural exclusion of "increasingly important spatial interdependencies as well as the conflictual nature of spatial formations" (Blotevogel 1996, p. 13).

An alternative discourse to traditional cultural landscape research has formed in North American landscape research since the 1950s with J. B. Jackson through his focus on vernacular landscapes – Jackson (1997, p. 343) places the vernacular landscape in the focus of his considerations: "It is a temporary product of much sweat and hardship and earnest thought." Thus, he not only differentiates himself from an aesthetic view of landscape, he also shares a rejection of the ideas of a harmonious and ultimately stable 'organism'. Instead of the search for a harmonious landscape structure of nature and culture, he focuses on the changeability of a landscape that is understood materially. He pleads for not idealizing Landscape One of the Middle Ages or Landscape Two of the Renaissance, also because of its simple comprehensibility, but for openly confronting Landscape Three of modernity with its complex connections (Jackson 1984). To some extent though, Jackson's approach was first received more intensively in German-language landscape research in the first decade of the twenty-first century (in the course of the international opening mentioned above; e.g. in the case of Franzen and Krebs 2005; Prominski 2004). But Jackson's broader approach, with its 'Landscape Three', was able to gain a popularity in the German-speaking world equivalent with that which the concept of the 'Zwischenstadt' by Thomas Sieverts' was able to achieve (1998 [1997]). It was a book that dealt with those spaces that were neither unambiguously city nor country, neither unambiguously nature nor culture, and that ran counter to the striving for uniqueness of the hegemonic discourse of the 'European city', which normatively strived for the dichotomous separation of city and country. Here Sieverts called for an 'open' approach to the 'Zwischenstädte' (intermediate cities; more detailed also at Hauser and Kamleithner 2006; Sieverts 2004; Vicenzotti 2019).

A high point of criticism of the Berkeley School was reached in Anglo-Saxon countries with the development of the 'new cultural geography' (Cosgrove 1989; Cosgrove and Jackson 1987; Duncan 1990). Although the new cultural geography also deals with historical contexts (admittedly contextualised and theoretically framed), it now focuses specifically on social aspects in an approach that integrates the urban. Furthermore, the 'new cultural geography' is "interested in the contingent nature of culture, in dominant ideologies and in forms of resistance to them" (Cosgrove and Jackson 1987, p. 95). Here the change from an essentialist basic attitude, in a rudimentary combination with empirical approaches, to a social constructivist dominated orientation is carried out. So, for Cosgrove (1984) is 'landscape not simply the world we see; it is a way of seeing the world'. In the 'new cultural geography', 'landscape' always remains present; it is conceptualised as 'text' or as an everyday construction (among other sources Denser 2019; Duncan 1990; Duncan and Duncan 1988). In Germany, on the other hand, the transition from traditional landscape geography to a 'modern' geography took place in the form of a break. The 'landscape paradigm' of geography was denied sustainability by younger researchers and students at the Kiel Geographers' Day in 1969: "The lack of social relevance of the subject, the arbitrariness and lack of ideology were criticised as ideology. The supposedly non-political, but ultimately conservative to reactionary, restorative geography was perceived as a discipline that blatantly violated the standards of conceptual debate achieved in neighbouring disciplines of the humanities and social sciences" (Gebhardt 2016, p. 45). With its "all too simple realism" (Kaufmann 2005, p. 102), it was largely replaced without resistance by a positivist-empiristic paradigm. This in turn had very different consequences for the usability of the word 'landscape' in physical and human geography: while in "mainstream anthropogeography it was not very career-promoting to speak of landscape" (Schenk 2006, p. 17). 'Landscape', thought of as a material object and mostly without explicit definition, was reused in natural science oriented physical geography in connection with ecosystem approaches as geoecology and landscape ecology (e.g. Eisel 2009; Kirchhoff and Trepl 2009a). The 'quantitative revolution' (Arnreiter and Weichhart 1998; Glasze 2015, p. 27) brought a far-reaching hegemony of positivist approaches in German geography, the category 'space' largely replaced the thematisation of 'landscape', and on a medium scale the concept of 'region' made a career, which was considered to be capable of being framed positivistically, unlike the 'essentialistically' burdened concept of landscape (Chilla et al. 2016). However, this farewell to 'landscape' was associated with unintended side effects: With the farewell to the exploration of the thematic field of 'landscape', "human geography has almost completely abandoned a field that used to be part of its core competence" (Gebhardt 2016, p. 48), with the result that, on the one hand, "a structure of tradition has been torn down which today's generation of teachers and students can no longer tie in with" (Gebhardt 2016, p. 49), and, on the other hand, the social need for competences in landscape analysis, especially in planning, is now being met by other disciplines (from biology to archaeology). After 1969, human geography remained a refuge for the study of landscape: historical geography. It deals specifically with the material (partly immaterial) heritage of landscape, which it primarily understood as 'historically grown cultural landscape' as representational (Fehn 1976; Harvey et al. 2018; Schenk et al. 1997; Schenk 2001a, 2006, 2011). The focus on heritage, in turn, has been increasingly developed in Anglo-Saxon geography since the '00s (Harvey 2001).

In discourse theoretical terminology, for German Human Geography the replacement of a hegemonic discourse by a framework (Weber 2018a; see Sect. 2.1), while Anglo-Saxon geography shows a more continuous development of landscape research. A further difference can be found in another theoretical perspective: In Anglo-Saxon geography (especially North American geography), a strong Marxist geography developed, which in part is connected with the 'new cultural geography' (for example, in the case of the Cosgrove and Daniels 1988) perspective (such as Harvey 1996). Here, for example, the concept of landscape was used in the form of criticism of unequal possibilities for the material manifestation of one's own interests, but also of the development and perpetuation of landscape-aesthetic ideas (Disneyfication: Warren 1994), which contribute to giving economic inequalities (such as the unequal distribution of property in a region) the appearance of legitimacy (Bermingham 1989; Cosgrove 1998). In West German geography, however, the response to Marxist ideas remained subdued for a long time, only in the 1990s, after 'reunification' did 'critical geography' strengthen, but without using the concept of 'landscape'. At about the same time, interest in the new cultural geography (including Gebhardt 2016) also awakened in Germany.

Despite the internationalisation of German-language geography, there are some specific features (especially with regard to landscape) today: The attention to phenomenological approaches in Anglo-Saxon landscape research in its German-language counterpart only takes place in rudimentary form (in particular with regard to Kazig 2013, 2019a, 2019b), the same is valid for the actors-network and the assemblage-theory (Kühne 2019e; Poerting and Marquardt 2019). While Anglo-Saxon landscape-related discourse research is more strongly based on the approaches of Michel Foucault's (1977), those in the German-speaking countries are dominated more with a connection to Laclau and Mouffe. On the other hand, a social constructivist approach has proved to be widespread in German-language landscape research on human geography, although it is more closely tied to sociological theory than to Anglo-Saxon geography. The autopoietic systems theoretical perspective is also more widespread in the German-speaking world than in the Anglo-Saxon world. Approaches to 'Critical Landscape Research' are more influenced by Bourdieu's sociology than of a Marxist orientation (more detailed: Kühne 2018f, 2019c).

The strong presence of constructivist approaches in landscape research has made it seem opportune to distinguish them as 'New Landscape Geography' (Kühne et al. 2018), in a geographical environment that still has difficulty with the concept of landscape. However, the still rather marginal position of landscape research in human geography in the German-speaking world has led to a far-reaching mutual tolerance of perspectives, which has developed into 'neopragmatic' combinations of different theoretical positions, if the object of research suggests it (Chilla et al. 2015; Eckardt 2014; Kühne 2019e).

4.6   Bottom Line

In summary, due to the multitude of connotations that have been laid out in the history of concepts, 'landscape' eludes a generally valid scientific definition. Nevertheless, Hard (1970c) is not to be followed, therefore the term is to be used only colloquially. Instead, the heuristic, research-conceptual, and that pertinent to spatial planning should be used (Kühne 2018b; Tillmann 2016), which implies that 'landscape' is a historically diverse carrier of meaning and thus offers manifold points of connection to everyday world understandings such as scientific approaches of different disciplines, paradigms, and theories. The prerequisite for this should be the insight that *the* landscape, *the* 'generic term' (Jessel 1995) or *the* 'super paradigm' (Hard 2002a) for 'landscape' does not and cannot exist. Nor can there be *the* science of landscape, since each discipline constructs its own understandings of landscapes (Trepl 1996). Landscape scientists and researchers are therefore invited to reflect on their disciplinary understanding of the landscape and how it is approached (Kühne 2019c; Schenk 2017) as well as interdisciplinary and trans-disciplinary cooperation (Berr 2018b). The following chapter shows what this means and to what extent such cooperation has been sought and implemented so far.

Landscape Research in Its Interdisciplinary and Transdisciplinary Challenge

5.1 Introductory Remarks

Contemporary landscape research is confronted with the differentiation of different landscape disciplines that address specific aspects of the phenomenon or concept of 'landscape' using specific paradigms, interpretative approaches and methods (see Berr and Schenk 2019). In view of the "complexity of reality" and the "pragmatically determined goals of knowledge" of the respective disciplines, such a differentiation from a historical and theoretical perspective is fundamentally "appropriate" from a factual and methodological point of view (Tetens 1999, p. 1769; see also Kühne 2019e) and unproblematic. Landscape research, however, finds itself in a doubly ambivalent situation due to these developments. In terms of research pragmatism, the plurality of disciplinary thematisations is an asset as far as the diversity of possible everyday language and scientific aspects and perspectives can be grasped and taken into account (Kühne 2018f, 2019c; Trepl 1996). But the corresponding research results often remain unrelated or are only 'combined' additively (Balsiger 2005, p. 214) without relating or organising the results to common research perspectives or application contexts. In a pragmatic application, the ambiguity, heterogeneity, and openness of meaning of the landscape concept – for example in the sense of a *'boundary object' (cf.* Gießmann and Taha 2017) – may provide "various identification possibilities" (Kühne 2018f, p. 324) for cooperation between scientists and non-experts, thus making it easier to link scientific discourses to everyday discourses (Kühne 2018f). At the same time, this exposes landscape research to the accusation that 'landscape' thus becomes a term "with high sympathy value and low risk of definition" (Franzen and Krebs 2005, p. 25). Interdisciplinary and transdisciplinary approaches in landscape research face this double ambivalence.

© Springer Fachmedien Wiesbaden GmbH, ein Teil von Springer Nature 2020
K. Berr and O. Kühne, *"Und das ungeheure Bild der Landschaft…"*, RaumFragen:
Stadt – Region – Landschaft, https://doi.org/10.1007/978-3-658-27756-7_5

In the following, we first illuminate disciplinary and cross-disciplinary landscape research in order to differentiate interdisciplinary and subsequently transdisciplinary approaches before concluding with a brief summary.

5.2   Disciplinary and Interdisciplinary Landscape Research

Scientific disciplines in their differentiated diversity are not, in the end, a reaction to the fact that the realms of reality to be researched are not accessible to individual scientists or research groups in a supposed 'totality', but only in scientifically specific characteristics or aspects, which accordingly require specific concepts, theories, methods, terminologies and, of course, scientific criteria (Defila and Di Giulio 1998, pp. 111–113). This "important basic methodological fact of any research" (Tetens 1999, p. 1769) is, as it were, the 'engine' of a dynamic that constantly produces new disciplines, brings them together, merges them into other disciplines or allows them to disappear. 'Disciplines' are therefore, on the one hand, included within "*historically grown* boundaries" (Mittelstraß 2005, p. 19). On the other hand, they represent a "cognitive and social unity within science" (Defila and Di Giulio 1998, p. 112) insofar as scientists join to form 'scientific communities' with specific paradigms, theories, research problems, and career paths (Defila and Di Giulio 1998).

Little attention is often paid to the fact that the term "discipline" indicates "a unitary point of view", "under which certain knowledge can be separated from others in terms of form". Regardless of whether a discipline is understood as a "historically grown summary" or as a "systematic-reconstructive compilation of knowledge and forms of knowledge", in order to define and delimit such 'units' in the sciences, criteria must be given "whereby the respective 'unity' of the discipline in question could be proved" (Gutmann 2005, p. 70). It should also be remembered that, by being included in "historically grown borders" (Mittelstraß 2005, p. 19), disciplines owe their origins to the problems of the living environment (cf. exemplary: Janich 2015, Gethmann 2010, 2011) and the disciplinary shifts of responsibility within the history of science (is 'heat' an 'object' of physics or chemistry; is 'landscape' an object of landscape planning, geography, or other disciplines?). This means that it is not objects, phenomena, or methods *alone* that "define the discipline, but the way in which we deal with them theoretically" (Mittelstraß 2005, p. 19). Disciplines are therefore '*constructs*' under pragmatic aspects of unity. Inevitably, therefore, "certain problems also elude the access of a single discipline" (Mittelstraß 2005, p. 19) and thus refer by themselves to the need for interdisciplinary research. Conversely, discipline-specific knowledge is always already dependent on knowledge created within other disciplines and is basically "*interdisciplinary* in *form*" (Gutmann 2005, p. 70). The knowledge acquired by disciplines can be used, as with Heinz Heckhausen (1987), and can also be described as a "theoretical level of integration", "at which the chosen object aspect is theoretically integrated, even reconstructed" (quoted after Balsiger 2005, p. 146).

This finding applies in particular to landscape-related disciplines such as spatial planning, geography, landscape planning, and landscape architecture, which, in comparison to 'compact' (Toulmin 1978) or 'hard' (Hard 2003) disciplines such as physics or chemistry "cannot fall back on an independent theoretical framework", "but rather as polyparadigmatic disciplines" (Dettmar 2018, p. 26 with reference to landscape architecture) must rely heavily on the concepts, methods, and results of other sciences. In addition, landscape-related disciplines, but also linguistics, literary studies, historical sciences, philosophy, art history, ethnology, and other sciences, in which "at every turn pre-scientific and scientific ideas and theory" are combined, are also represented (Hard 2003, p. 180) and always react to problems in everyday life that "owe their existence to questions other than those of science alone" (Mittelstraß 2005, p. 19). They share their 'objects' and main terms with the pre-scientific understanding of everyday life and are dependent on their ability to connect with negotiation processes and the motivations, convictions, and values of different everyday actors.

This already outlines the first indications of functions, tasks, and forms of interdisciplinary landscape research. One form is concerned with boundaries and separations between different scientific disciplines and attempts to overcome them. The other form is concerned with boundaries and separations between scientific disciplines and the everyday world and attempts to overcome them.

5.3 Interdisciplinary Landscape Research

The diversity of different thought styles, paradigms, theories, and concepts – which are simultaneously given with the diversity of landscape-related disciplines and which each address a specific aspect or research problem – has been and is regarded by many scientists as an obstacle to knowledge and research. Therefore, a cooperation of the disciplines and their scientists was and is aimed at an effort to arrive at joint research solutions. However, for a long time there have been complaints about the "incooperativeness of the specialist sciences" (Lorenzen 1974, p. 135). The sociologist Ralf Dahrendorf has described this 'incooperativeness' of many scientists as 'territorial behaviour', inasmuch as "every researcher in a way cuts his own small garden out of the area of scientific problems presented as an endless piece of land, in order to fence it in as quickly as possible and not let anyone else in" (Dahrendorf 1972, p. 304) – in other words, like in an allotment garden colony (Homann and Suchanek 2005, p. 393). Joint research perspectives or application contexts can then hardly be organised. For decades, forms of possible cooperation between representatives of different disciplines have therefore been the subject of controversial discussion, with keywords such as 'multidisciplinarity', 'polyperspectivity' and, in particular, the model of 'interdisciplinarity' (cf. Balsiger 2005, pp. 157–173; Grunwald and Schmidt 2005; Jungert et al. 2010; Kocka 1987).

Landscape research has also responded to the problem of fragmentation of scientific knowledge (Eisel 1992b, c; Hard 2002b; Kühne 2008b, 2018f; Trepl 1996). Attempts, within the framework of a conceptual unification (cf. Jessel 1995, p. 7), to develop an all-encompassing generic term that integrates all connotations or a universal and consistent unifying generic term – or a "super paradigm" (Hard 2002a, p. 67) – and thus to "define landscape or cultural landscape generally valid" seem to many authors to be suspect or "illusory" (Leibenath and Gailing 2012, p. 58). It is also questionable whether such a conceptual standardization can be meaningful at all. Such a unification would correspond to an understanding of science that Ulrich Eisel (1992a) described as 'euphoric interdisciplinarity', namely, if landscape research would strive for a 'whole' as 'unity of the world' or as 'unity of the object (and knowledge)' (Eisel 1992a, p. 16). This would once again take up a Romantic figure of thought in the form of a 'holistic' understanding of science (cf. Sect. 3.5). Behind such a striving for "unity instead of multiplicity" (Gethmann 1991, p. 352) or for a "cognitive unification of the sciences into a unified science" is ultimately an "ideological scientism" (Gethmann 1991, p. 350). Demands for such 'unification' are therefore owed to an often unacknowledged 'scientific ideological postulate' (Balsiger 2005, p. 247), an 'ideology of interdisciplinarity itself' (Eisel 1992a, p. 7) or the longing for a holistic unity within the framework of a 'discourse of transcendence' (Klein 2014, p. 70). Against such ideologies it can therefore be demanded not to seek or construct a "whole" as "unity of the world" or as "unity of the object (and knowledge)", but to participate in a "reasonable organization of dissent" (Eisel 1992a, p. 16). Interdisciplinarity is then, if it is to succeed, an "object-constitutive discourse on differences": For not "the unity of the object (and knowledge) must be 'sought', but the difference of the theories about the object must be 'determined'. This difference is the unity of the object as interdisciplinary" (Eisel 1992a, p. 8), i.e. a 'differentiated unity' (Trepl 1996) as a unity of different perspectives and thematic aspects.

In landscape research, at least two typical forms of nonholistic answers to the fragmentation of knowledge can be distinguished. First, a negative understanding of 'multidisciplinarity' in the sense of merely additive unification (Balsiger 2005, p. 214) of differentiated knowledge, as opposed to a positively modelled "multidisciplinarity" (Potthast 2010, p. 180), which can be described using Kühne as a "multi-perspective approach" (2018f, p. 2) in order to "provide an overview of the main strands of landscape research" (2018f, p. 3). This form of interdisciplinary landscape research can be understood and handled as a form of pragmatic systematisation of knowledge with an educational and orientation function. In this sense, "points of view can be grouped and related to other positions" – such an approach can "order the existing views on the concept of landscape and facilitate discourse on similarities and differences" (Hokema 2018, p. 30). For example, landscape terms can be arranged according to language or reflection levels (Leibenath and Gailing 2012; Schenk 2017) or different forms of ideal-typically systematizations and reconstructions developed (Hartlieb von Wallthor and Quirin 1977; Jackson 1984; Jessel 2005; Kirchhoff and Trepl 2009b; Kühne 2006, 2018f; Prominski 2004; Trepl 2012; Vicenzotti 2011). The diversity of different landscape concepts and theories can then, according to a proposal going beyond this, be positively seen as an offer to the scientifically fruitful "contradiction" and to the "productive friction"

(Leibenath and Gailing 2012, p. 59) of the scientists as interpreted among themselves. By reflecting on the meaning and use of concepts and theories, it is expected that an "increase in the conceptual security of designers and planners" can and should be achieved (Vicenzotti 2012, p. 272).

Secondly, and this corresponds to an understanding of interdisciplinarity that aims at cooperation between different landscape researchers and not at a multidisciplinary synopsis of knowledge by individual landscape researchers, there have been and still are interdisciplinary joint conferences, the results of which are published in corresponding research volumes (e.g. Felten et al. 2012; Hartlieb von Wallthor and Quirin 1977; Krebs and Seifert 2012).

Regardless of these multidisciplinary and interdisciplinary efforts, however, the long-standing scientific interest in 'landscape' does not lead to a convergence of views; rather, it has developed different discourses with their own discourse sovereignties and logics, some of which are clearly transversal to disciplinary approaches. By Kühne (2006; Kühne 2008b, d, 2018f; see also Hupke 2015; Jones and Daugstad 1997; Wojtkiewicz and Heiland 2012), four discourses are currently dominant:

1. The discourse of the preservation and restoration of a physical space understood as a 'cultural landscape' – a discourse based on an essentialist, partly positivist interpretation of landscape, especially in landscape planning, historical geography, monument conservation, and other disciplines (see Hage and Bäumer 2019; Hartz 2019; Heiland 2019; Weber and Kühne 2019b).
2. The discourse of the successionist development of a space understood as 'landscape' is based normatively on the structuring of the physical foundations of appropriated physical landscape as a secondary consequence of ecological or social developments. 'Landscape' is not assigned here, as in the previous approach, its own value arising from its 'essence'. It is ultimately understood as a registration plate for natural and social developments. This approach, which is based on a positivist understanding of science, is particulary represented by biologists on the one hand and economists on the other (Leser 2019a, b; Weber and Kühne 2019b).
3. The discourse of the reflexive design of a physical space understood as 'landscape' is shaped by the notion of developing new interpretations and evaluations of the individual and social construction of landscape through targeted (artistic) interventions in the material world. This view – which is based on a social constructivist understanding of the world – is especially represented in the disciplines of fine arts, social sciences, and landscape architecture (see Berr 2019d; Hennecke 2019).
4. The discourse of the reinterpretation of social landscape concepts pursues the goal of changing landscape understandings without interfering with the material world through adaptations. This view, which is based on constructivist approaches, is frequently represented by social scientists and humanities scholars (Berr et al. 2019; Berr and Kühne 2019a; Denzer 2019; Hengst 2019; Kühne 2019f; Weber 2019b; Weber and Kühne 2019a).

These discourses – which break through disciplinary boundaries – are characterised by competition for interpretive sovereignty, which in turn can trigger clear differences within the disciplines involved, and are at the same time associated with discourse coalitions in an interdisciplinary context. For example, essentialistically interpretable approaches to wind power are seen as 'landscape disfigurement' (Nohl 2016) discursive theory, in which different attributions to 'landscape' are focused beyond an irrefutable basis (Kühne 2019c; Leibenath and Otto 2012; Otto and Leibenath 2013; Weber 2017, 2018a). This struggle for interpretive sovereignty, which is ultimately based on a struggle for academic reputation, scientific reputation, and financial resources, is thus always associated with the danger of unreflected or methodically uncontrolled reductionisms, essentialisms, and dogmatic positions. In such a 'struggle for recognition' (Honneth 1992; Eisel 1997), a struggle for collegial 'attention' and institutional security ultimately manifests itself in the current 'scientific enterprise' (Franck 2007).

The modern development of the sciences can be reconstructed as the background of such struggles for recognition and sovereignty of interpretation, the character of which – as early as 1938 – Martin Heidegger characterised as "enterprise" and therefore spoke of the "enterprise character of research" and the "institutional character of the sciences" (Heidegger 1963 [1938], p. 77–78). This "operational character" can be seen, for example, in "promoting the alternate review and communication of results" (Heidegger 1963 [1938], p. 78) – which then inevitably leads to the struggle for "attention" within the framework of an "attention economy" (Franck 2007; cf. in detail for landscape studies: Kühne 2008b, p. 270–281). Now "science as a profession" (Weber 2011 [1919]) can be understood and practiced by 'professionals'. The sciences develop specific career paths, they are increasingly developed into more and more differentiated disciplines, subjects, and research directions, this leads to a specialisation and an incessant need for innovation.

If applied positively and *objectively,* every science or discipline is also a struggle for the *best possible* scientific explanation of questionable phenomena within the framework of concepts and theories. However, all research is subject to 'fallibilism' (Popper 1963), so no one can claim scientifically proven knowledge about which concepts, theories, or paradigms offer the *undoubtedly* 'right' explanations or solutions with regard to questionable phenomena, pressing problems, or specific research questions. In order to keep open the horizon of possible explanations and proposed solutions against immunization strategies or presumptuous interpretative sovereignties, this diversity of theories and concepts is scientifically productive, insofar as the open horizon of this plurality requires the advocates of concepts and theories to intensify their explanatory work and thus generally increases the chances of generating the best possible theories and corresponding knowledge at human discretion. In the economic market this means: 'Competition stimulates business'; this certainly also applies to the scientific reasoning business. For this reason, Dahrendorf expresses therefore "the mutual criticism of the researcher's condition of the possibility of avoiding the dogmatised error. Such criticism demands of the individual, above all, openness to new and better solutions to 'his' problems" (1972, p. 305). Here, as it were, an "ethics of freedom" also shows itself in the form of an "ethics of the [scientific; KB and OK] conflict, of the endured and tamed antagonism" (Dahrendorf 1972, p. 313–314). Therefore, a choice of theories cannot be directed *against* other theories with a sole claim to explanation, but can only be proposed in accordance with clearly defined research purposes as *a* solution with specific claims to explanation in addition to theories with other claims to explanation.

In general, the accusation was made against interdisciplinary research that disciplines would be marginalised, their knowledge and methods replaced or abolished by interdisciplinary knowledge. The response to this accusation was the reference to the complementarity of disciplinarity and interdisciplinarity, insofar as a cross-disciplinary research process always builds on the factual and methodological knowledge of the respective disciplines responsible for the specific problem. Interdisciplinary research can therefore "always only be understood as 'complementary' […] to a disciplinary research process" (Balsiger 2005, p. 189). It is precisely the "diversity of perspectives" compared with a "specialised focus" of the individual disciplines that distinguishes this form of research (ibid.). "Good disciplinarity" is therefore "a prerequisite for good interdisciplinarity" (Sukopp 2010, p. 19) and therefore remain "the institutional organizational form of the sciences" (Mittelstraß 1998, p. 45).

5.4 Transdisciplinary Landscape Research

Landscape-related disciplines do not produce theoretical knowledge for their own sake, but are also confronted with non-scientific challenges, such as the needs, interests, demands, user expectations, and conflicts of actual living world actors. Demographic change, urbanization and re-urbanisation processes, the energy revolution, new private and public demands for use in cities and rural areas are creating new economic, social, cultural, ecological, and political challenges for landscape practice and research. Practical cooperation is required not only between scientists – as in the case of interdisciplinary cooperation – but also on the basis of the stakeholder model of business ethics (Freeman 1984) – stakeholders from politics, administration, business, and the so-called 'population' must also be involved. The actors involved should no longer be regarded as 'objects' in a research perspective, but also as active participants ('subjects') in negotiation or participation processes.

In view of the increasing efforts of citizens to participate, politics and administration are under increasing political pressure to legitimise themselves (Walter et al. 2013). Management exclusively 'from above' ('*top down*'), which ignores subsystem logics, different actor constellations, the influence of the mass media and social networks, threatens political-administrative ineffectiveness and democratic unacceptance. In recent decades, science and practice have reacted to these developments with the concept of governance (Benz 2004; Gailing 2018; cf. exemplary: Kühne 2018f, pp. 303–319; Leibenath and Lintz 2018; Mayntz 2005; Weber et al. 2018, pp. 30–35; Zürn 2008).

As expressed by Arthur Benz (2004, binding), governance is understood to mean "new forms of social, economic and political regulation, coordination and control in complex institutional structures in which state and private actors usually work together". The concept of transdisciplinarity can also be assigned to this problem context and has also been discussed for a long time without this term itself having been explicitly used (cf. Balsiger 2005; Berr 2018a; Grunwald and Schmidt 2005; Jahn 2008; Sukopp 2010; Vilsmaier and Lang 2014 and others).

It was also questionable how interdisciplinary and transdisciplinarity could be precisely distinguished. Symptomatic of this problem is a manifestation of Mittelstraß (2005, p. 19): "Interdisciplinarity in the rightly understood sense […] is in truth transdisciplinarity". Seen in this light, transdisciplinarity would be the elaborate version of interdisciplinarity in need of improvement and would therefore "merely describe a particularly high degree of intensity of interdisciplinary cooperation" (Laitko 2018, p. 11). Meanwhile in corresponding publications by many authors (Balsiger 2005; Grunwald and Schmidt 2005; Jahn 2008; Jungert et al. 2010; Sukopp 2010; Völker 2004) interdisciplinarity is used as a term for overcoming or crossing disciplinary boundaries and separations, while transdisciplinarity is employed as a concept for the "thesis of the partial delimitability of science and society" (Grunwald and Schmidt 2005, pp. 7–8). Transdisciplinarity is therefore distinguished from "interdisciplinarity in the narrower sense" by "the fact that explicitly science-external questions and persons actively participate in research" (Potthast 2010, p. 180–181). In other words: Transdisciplinarity can be "as an actor-extended variation" (Weith and Danielzyk 2016, p. 10) defined by interdisciplinarity. Similar provisions can be found, for example, in Jahn (2008), Laitko (2018), Pohl and Hirsch Hadorn (2008) or Vilsmaier and Lang (2014). Goals are to gain an "argumentative unity" (Mittelstraß 2005, p. 23) of scientific and life-world arguments, thereby avoiding uncritical, unreflected, or even dogmatic insistence on sovereignty of interpretation (Gieryn 1983; Pohl and Hirsch Hadorn 2008), instead the "different decision logics and rules" (Loibl 2005, p. 34) to allow and understand disciplinary and living world actors and thus the diversity of points of view (Pohl and Hirsch Hadorn 2008, pp. 13–14) and thus to form the foundation "for a general understanding of the problem, for a shared language, for the need for action to be formulated and for the development of a concrete goal and action perspective" (Weith and Danielzyk 2016, p. 10).

The varieties of transdisciplinary landscape research in a broad sense can currently be found in environmental and sustainability research in German-speaking countries (cf. Pohl and Hirsch Hadorn 2008; Schäfer 2013; Vilsmaier and Lang 2014; Waag 2012) and nowadays is also taking place in architectural science (cf. Doucet and Janssens 2011; Hauser and Weber 2015) and the spatial sciences (cf. Weith and Danielzyk 2016). Except for individual works and research projects (cf. notes by Dettmar 2018; Schmidt et al. 2018; Stemmer 2016; Tress et al. 2003; Tress and Tress 2001; Weber, Kühne et al. 2016; Weber and Kühne 2016) there is hardly any *systematic* interest in transdisciplinary research for landscape research yet (see Berr 2005; Küster 2012b; Weith and Danielzyk 2016).

5.5   Conclusion and Outlook

To research means to work regularly, or actually permanently, at boundaries and cross them. This is also particularly evident in the context of 'landscape' oriented research, which regularly reaches discipline-related limits. For example, dealing with the expansion of the electricity grid in Germany and with 'landscape-related arguments' brings together spatial, planning, control-related, aesthetic, technical, etc., aspects. These are aspects that are difficult to deal with in a narrow disciplinary context – whatever the concrete context might be. At the same time, we are regularly confronted with the considerations of Ralf Dahrendorf (1972, p. 304), who wants to stake out 'his little garden' in such a way that it could not be entered by others at all. Perhaps, however, this is less a question of discipline-related boundaries than of a common orientation of 'research circles'. Getting involved in interdisciplinary landscape research, with different theoretical and empirical approaches, happens already in the basic approach, but only in direct confrontation with 'specific' other researchers the challenges become clear under certain circumstances, including the finding of a common 'language' and perspective, e.g. in such a network development with the participation of landscape architects, communication scientists, political scientists, and geographers (starting from the respective subject; Weber, Kühne et al. 2016). In many cases, however, scientists move out of the 'comfort zone' only 'really' when they enter into 'concrete practice', when actors with an application-related focus (should) be integrated transdisciplinarily. A lack of mutual understanding can become a 'drama' but can also be an enrichment with a great deal of effort in building common foundations. The 'landscape' offers manifold potential; in any case, resulting from its ability to be connected to everyday life and research, but this potential has not been exploited to any great extent to date.

Bottom Line

6

This book focuses on two of the three components of 'threefold landscape transformation': the societal, here in relation to the common sense understanding of landscape (in Chap. 3), and the individual, in relation to influencing and modifying the scientific understanding of landscape by individuals with 'expert knowledge' (Chap. 4). Here the social distribution of power (which is discussed more intensively elsewhere; Kühne 2008b, c, 2015b, 2018d, g) becomes clear as a result of differentiation: There are people in society who are in a position to discursively anchor social reinterpretation options of landscape; others cannot, on the contrary, as they are threatened with the loss of social recognition. Innovations in interpretation from the context of expert communication then diffuse – also by means of systematic socialisation (specifically of the school) – into the common sense understanding of landscape (where they are then far less or no longer terminologically defined). Edmund Husserl (1859–1938) thematised this phenomenon that scientific ideas, concepts, and theories can flow into the pre-scientific 'natural attitude' of everyday understanding, under the title 'influx' into the everyday world (Husserl 1954, p. 141; cf. Held 1991, p. 106). This means that the living world as a world of natural attitudes does not simply face the historical, cultural, and scientific world. Rather, this living world is "a world that is historically enriched by the practice taking place in it and its sedimentation, by the influx. It's the concrete historical world. In this historically evolving world of natural attitudes, the results of philosophical-scientific thinking, which rises above the first natural attitude, are also incorporated" (cf. Held 1991, p. 107). As a result, (first) the dichotomy between everyday landscape construction by laypersons and systemic construction by experts hybridises as a result of educational expansion, (second) but also through strategic use of scientific results of landscape research in conflicts over landscape becomes a medium of science (among many: Berr et al. 2019; Berr and Kühne 2019b; Bues 2019; Kühne 2018a, 2019a). This also means, however, that (thirdly) scientific communication aims to generate

K. Berr and O. Kühne, *"Und das ungeheure Bild der Landschaft…"*, RaumFragen: Stadt – Region – Landschaft, https://doi.org/10.1007/978-3-658-27756-7_6

new patterns of landscape interpretation and assessment (others speak of scientific progress) in order to achieve distinctive gains in comparison with an expanding common sense understanding of landscape (in principle, for this purpose: Bourdieu 1982, 2016; Bourdieu and Passeron 1973; Diaz-Bone 2010 in relation to landscape: Kühne 2008c, 2018d).

The common sense understanding of landscape – as shown – is also characterised – in comparison to other European understandings – by a strong territorial reference, which is already established in the Middle Ages, but it is also provided with a strong emotional component of belonging. Set in the Romantic period and turned into a private life practice in the Biedermeier period, home and landscape were provided with an amalgamated intersection (Kühne 2009b) which continues to have an impact in numerous conflicts over landscape changes to this day (Berr 2019c; Berr and Kühne 2019b; Hoeft et al. 2017; Kühne 2018a, 2019g; Kühne et al. 2019; Weber et al. 2017, 2018; Weber and Jenal 2018). On the basis of this amalgamation of landscape and 'Heimat' this connection could be politicised and served – essentialistically interpreted – as a fundamental justification model for the *Heimatschutz* movement, which represents the origin of the nature conservation movement in Germany. A particular focus of the conservation of preferred material objects over the physical manifestos of modernization (especially industrialization) was the (German) forest, which in the Romanesque period, and still today, functions as a physical element of the German national myth. Not only in the common sense understanding of landscape, but also in the scientific discussion about landscape there are some specifics. First of all, the conceptual version of landscape at the end of the nineteenth and beginning of the twentieth century was strongly influenced by the then shared understanding: landscape was simultaneously region, territory, and aesthetic synopsis, with a 'being' behind it. An understanding that made an international career in the scientific community and remained at its core until the 1960 s – despite the integration of increasing empirical approaches based on a positivist worldview. However, while Anglo-Saxon geography succeeded in translating the theme of landscape into current scientific perspectives, such as positivist, (neo)Marxist, and not least Constructivist as well, in recent years, non-representational ones, the end of 'traditional landscape geography' in German human geography meant a break with the theme of landscape. Geographical landscape research has not been able to recover from this rupture to this day, not even with a much differentiated constructivist diversity of perspectives.

The variety of understandings and approaches becomes clear in the reflection of the explanations on the German-language landscape understandings, their genesis in common usage and science, as well in the context of the comparison to non-German-language understandings. With regard to scientific understanding, the value of the plurality of different scientific worldviews for the appropriateness of investigations can be promoted – also from the experience of the development of German-speaking geography. A hegemonic perspective can hardly grasp the differentiation and complexity of the theme 'landscape'. In this respect, a fundamental superordination and subordination of theoretical approaches seems questionable; depending on concrete questions, positivist as well as constructivist approaches – also approaches that go beyond representatively – can provide insights, and even essentialist approaches, provided they are subject to reflexive framing, can enrich the diversity of perspectives on landscape. The same applies to the different culturally and linguistically bound patterns of landscape interpretation and assessment. They represent a resource for the respectful handling of 'landscape'.

References

Adorno, T. W. (1970a). *Ästhetische Theorie. Gesammelte Schriften Band 7.* Frankfurt a. M.: Suhrkamp.

Adorno, T. W. (1970b). *Gesammelte Schriften. Band 6. Negative Dialektik. Jargon der Eigentlichkeit.* Frankfurt a. M.: Suhrkamp.

Albert, H. (1960). Wissenschaft und Politik. In E. Topitsch (Ed.), *Probleme der Wissenschaftstheorie. Festschrift für Victor Kraft* (pp. 201–232). Berlin: Springer.

Albert, G. (2005). *Hermeneutischer Positivismus und dialektischer Essentialismus Vilfredo Paretos.* Wiesbaden: VS Verlag.

Andrews, M. (1989). *The Search for the picturesque. Landscape aesthetics and tourism in Britain, 1760–1800.* Stanford: Stanford University Press.

Antrop, M. (2018). A brief history of landscape research. In P. Howard, I. H. Thompson, E. Waterton, & M. Atha (Eds.), *The routledge companion to landscape studies* (2nd ed., pp. 1–16). Abingdon: Routledge.

Appleton, J. (1975). *The experience of landscape.* London: Wiley.

Appleton, J. H. (1980). *Landscape in the arts and the sciences.* Hull: University of Hull.

Aristoteles. (2009). *Politik. Neuausgabe* (Nach der Übersetzung von Franz Susemihl mit Einleitung, Bibliographie und zusätzlichen Anmerkungen von Wolfgang Kullmann). Reinbek bei Hamburg: Rowohlt.

Arnreiter, G., & Weichhart, P. (1998). Rivalisierende Paradigmen im Fach Geographie. In G. Schurz & P. Weingartner (Eds.), *Koexistenz rivalisierender Paradigmen. Eine post-kuhnsche Bestandsaufnahme zur Struktur gegenwärtiger Wissenschaft* (pp. 53–85). Opladen: Westdeutscher Verlag.

Aschenbrand, E. (2017). *Die Landschaft des Tourismus. Wie Landschaft von Reiseveranstaltern inszeniert und von Touristen konsumiert wird.* Wiesbaden: Springer VS.

Assunto, R. (1963). *Die Theorie des Schönen im Mittelalter.* Köln: DuMont.

Balsiger, P. W. (2005). *Transdisziplinarität. Systematisch-vergleichende Untersuchung disziplinenübergreifender Wissenschaftspraxis.* München: Fink.

Bätzing, W. (2000). Postmoderne Ästhetisierung von Natur versus „schöne Landschaft" als Ganzheitserfahrung – Von der Kompensation der „Einheit der Natur" zur Inszenierung von Natur als Erlebnis. In A. Arndt, K. Bal, & H. Ottmann (Eds.), *Hegels Ästhetik. Die Kunst der Politik - Die Politik der Kunst. Zweiter Teil* (pp. 196–202). Berlin: Akademie.

Bätzing, W. (2002). Verschwindet der ländliche Raum? Perspektiven nach 2001. *PRO REGIO, 26*(27), 5–11.

Bauch, K. (1957 [1937]). Anfänge der neuzeitlichen Kunst. In Joachim Jungius-Gesellschaft der Wissenschaften (Ed.), *Die Entfaltung der Wissenschaft. Zum Gedenken an Joachim*

© Springer Fachmedien Wiesbaden GmbH, ein Teil von Springer Nature 2020

K. Berr and O. Kühne, *"Und das ungeheure Bild der Landschaft…",* RaumFragen: Stadt – Region – Landschaft, https://doi.org/10.1007/978-3-658-27756-7

Jungius (1587-1657). Vorträge gehalten auf der Tagung der Joachim Jungius-Gesellschaft der Wissenschaften, Hamburg, am 31. Okt./1. Nov. 1957 aus Anlaß der 300. Wiederkehr des Todestages von Joachim Jungius (pp. 118–139). Hamburg: Augustin.

Baumüller, B., Kuder, U., & Zoglauer, T. (Eds.). (1997). *Inszenierte Natur. Landschaftskunst im 19. und 20. Jahrhundert.* Stuttgart: Deutsche Verlags-Anstalt.

Bausinger, H. (1984). Auf dem Wege zu einem neuen, aktiven Heimatverständnis. Begriffgeschichte als Problemgeschichte. In H.-G. Wehling (Ed.), *Heimat heute* (Kohlhammer-Taschenbücher Bürger im Staat, Vol. 1065, pp. 11–27). Stuttgart: Kohlhammer.

Bender, B. (1982). *Ästhetische Strukturen der literarischen Landschaftsbeschreibung in den Reisewerken des Fürsten Pückler-Muskau.* Frankfurt a. M.: Lang.

Benz, A. (Ed.). (2004). *Governance – Regieren in komplexen Regelsystemen. Eine Einführung.* Wiesbaden: VS Verlag.

Berger, P. L., & Luckmann, T. (1966). *The social construction of reality. A treatise in the sociology of knowledge.* New York: Anchor books.

Berr, K. (2005). Landschaft – Die Rehabilitierung des verschmähten Naturschönen in der Kunst. In U. Franke & A. Gethmann-Siefert (Eds.), *Kulturpolitik und Kunstgeschichte. Perspektiven der Hegelschen Ästhetik. Sonderheft der Zeitschrift für Ästhetik und Allgemeine Kunstwissenschaft* (pp. 119–142). Hamburg: Meiner.

Berr, K. (2008). Carus und Hegel über Landschaftsmalerei. Landschaftsästhetik nach dem „Ende" der Landschaftsmalerei. In A. Gethmann-Siefert & B. Collenberg-Plotnikov (Eds.), *Zwischen Philosophie und Kunstgeschichte. Beiträge zur Begründung der Kunstgeschichtsforschung bei Hegel und im Hegelianismus* (pp. 243–256). München: Fink.

Berr, K. (2009). *Hegels Bestimmung des Naturschönen. Zur Betrachtung und Darstellung schöner Natur und Landschaft.* Saarbrücken: Südwestdeutscher Verlag für Hochschulschriften.

Berr, K. (2013). Wahrheit und »Möglichkeitssinn« – Hegels Ästhetik im Kontext moderner Kultur. In H. Friesen & M. Wolf (Eds.), *Kunst, Ästhetik, Philosophie. Im Spannungsfeld der Disziplinen* (pp. 129–168). Münster: Mentis.

Berr, K. (2018a). Einführung. In K. Berr (Ed.), *Transdisziplinäre Landschaftsforschung. Grundlagen und Perspektiven* (pp. 1–12). Wiesbaden: Springer VS.

Berr, K. (Ed.). (2018b). *Transdisziplinäre Landschaftsforschung. Grundlagen und Perspektiven.* Wiesbaden: Springer VS.

Berr, K. (2019). Heimat und Landschaft im Streit der Weltanschauungen. In M. Hülz, O. Kühne, & F. Weber (Eds.), *Heimat. Ein vielfältiges Konstrukt* (pp. 27–51). Wiesbaden: Springer VS.

Berr, K. (2019). Klassiker der Landschaftsforschung und ihre gegenwärtige Wirkung. In O. Kühne, F. Weber, K. Berr, & C. Jenal (Eds.), *Handbuch Landschaft* (pp. 39–53). Wiesbaden: Springer VS.

Berr, K. (2019). Konflikt und Ethik. In K. Berr & C. Jenal (Eds.), *Landschaftskonflikte* (pp. 109–129). Wiesbaden: Springer VS.

Berr, K. (2019). Landschaftsarchitektur. In O. Kühne, F. Weber, K. Berr, & C. Jenal (Eds.), *Handbuch Landschaft* (pp. 231–244). Wiesbaden: Springer VS.

Berr, K., & Jenal, C. (Eds.). (2019). *Landschaftskonflikte.* Wiesbaden: Springer VS.

Berr, K., & Kühne, O. (2019). Moral und Ethik von Landschaft. In O. Kühne, F. Weber, K. Berr, & C. Jenal (Eds.), *Handbuch Landschaft* (pp. 351–365). Wiesbaden: Springer VS.

Berr, K., & Kühne, O. (2019). Werte und Werthaltungen in Landschaftskonflikten. In K. Berr & C. Jenal (Eds.), *Landschaftskonflikte* (pp. 65–88). Wiesbaden: Springer VS.

Berr, K., & Schenk, W. (2019). Begriffsgeschichte. In O. Kühne, F. Weber, K. Berr, & C. Jenal (Eds.), *Handbuch Landschaft* (pp. 23–38). Wiesbaden: Springer VS.

Berr, K., Jenal, C., Kühne, O., & Weber, F. (2019). Inter- und transdisziplinäre Landschaftsforschung. In O. Kühne, F. Weber, K. Berr, & C. Jenal (Eds.), *Handbuch Landschaft* (pp. 165–180). Wiesbaden: Springer VS.

Berr, K., Jenal, C., & Kindler, H. (2019). Landschaftskonflikte. In O. Kühne, F. Weber, K. Berr, & C. Jenal (Eds.), *Handbuch Landschaft* (pp. 367–382). Wiesbaden: Springer VS.

Bertels, L. (1997). *Die dreiteilige Großstadt als Heimat. Ein Szenarium.* Opladen: Leske + Budrich.

Birnbacher, D. (1992). Theodor W. Adorno: Negative Dialektik (1966). In Reclam (Ed.), *Interpretationen. Hauptwerke der Philosophie. 20. Jahrhundert* (pp. 335–361). Stuttgart: Reclam.

Blotevogel, H. H. (1996). Aufgaben und Probleme der Regionalen Geographie heute. Überlegungen zur Theorie der Landes- und Länderkunde anläßlich des Gründungskonzepts des Instituts für Länderkunde, Leipzig. *Berichte zur deutschen Landeskunde, 70*(1), 11–40.

Blumenberg, H. (1957). Nachahmung der Natur. Zur Vorgeschichte der Idee des schöpferischen Menschen. *Studium Generale, 10*(5), 266–283.

Blumenberg, H. (1984). *Der Prozeß der theoretischen Neugierde.* Frankfurt a. M.: Suhrkamp.

BNatSchG. (2009 [1976]). Gesetz über Naturschutz und Landschaftspflege (Bundesnaturschutzgesetz). https://www.gesetze-im-internet.de/bnatschg_2009. Accessed 17 May 2018.

Böheim, J. (1930). *Das Landschaftsgefühl des ausgehenden Mittelalters.* Leipzig: B. G. Teubner.

Böhme, H. (1995). Materialismus oder Konstruktivismus. Eine falsche Alternative – aus der Sicht der Goethezeit. In M. Großheim (Ed.), *Leib und Gefühl. Beiträge zur Anthropologie* (Lynkeus, Vol. 1, pp. 129–140). Berlin: Akademie.

Bollnow, O. F. (1941). *Das Wesen der Stimmungen.* Frankfurt a. M.: Klostermann.

Bondeli, M. (1990). *Hegel in Bern* (Hegel-Studien, Beiheft, Vol. 33). Bonn: Bouvier.

Born, M. (1977). *Geographie der ländlichen Siedlungen. 1: Die Genese der Siedlungsformen in Mitteleuropa.* Stuttgart: Teubner.

Borsche, T. (1996). Einleitung Sprachphilosophische Überlegungen zu einer Geschichte der Sprachphilosophie. In T. Borsche (Ed.), *Klassiker der Sprachphilosophie. Von Platon bis Noam Chomsky* (pp. 7–13). München: Beck.

Bourassa, S. C. (1991). *The aesthetics of landscape.* London: Belhaven Press.

Bourdieu, P. (1982). *Leçon sur la leçon.* Paris: Les Éditions de Minuit.

Bourdieu, P. (2005 [1983]). Politik, Bildung und Sprache. In P. Bourdieu (Ed.), *Die verborgenen Mechanismen der Macht* (pp. 13–30). Hamburg: VSA.

Bourdieu, P. (2016). *La distinction: Critique sociale du jugement* (Le Sens commun). Paris: Editions de Minuit.

Bourdieu, P., & Passeron, J.-C. (1973). *Grundlagen einer Theorie der symbolischen Gewalt.* Frankfurt a. M.: Suhrkamp.

Brühl, C. (Ed.). (1971). *Capitulare de villis. Cod. Guelf. 254 Helmst. der Herzog August Wilhelm Bibliothek Wolfenbüttel [Faksimile-Ausgabe].* Stuttgart: Müller und Schindler.

Bruns, D., & Kühne, O. (Eds.). (2013). *Landschaften: Theorie, Praxis und internationale Bezüge.* Schwerin: Oceano Verlag.

Bruns, D., Kühne, O., Schönwald, A., & Theile, S. (Eds.). (2015). *Landscape culture – Culturing landscapes. The differentiated construction of landscapes.* Wiesbaden: Springer VS.

Bues, A. (2019). Landschaftsargumente in Windkraftkonflikten: Ein internationaler Vergleich. In K. Berr & C. Jenal (Eds.), *Landschaftskonflikte* (pp. 411–422). Wiesbaden: Springer VS.

Burckhardt, J. (1976 [1859]). *Die Kultur der Renaissance in Italien. Ein Versuch.* Stuttgart: Kröner.

Burckhardt, L. (2004). Zwischen Flickwerk und Gesamtkonzeption (1982). In L. Burckhardt (Ed.), *Wer plant die Planung? Architektur, Politik und Mensch* (pp. 99–106). Berlin: Martin Schmitz Verlag.

Burckhardt, L. (2006). Landschaft (1998). In M. Ritter & M. Schmitz (Eds.), *Warum ist Landschaft schön? Die Spaziergangswissenschaft* (pp. 114–123). Kassel: Martin Schmitz Verlag.

Burckhardt, L. (2006). Natur ist unsichtbar (1989). In M. Ritter & M. Schmitz (Eds.), *Warum ist Landschaft schön? Die Spaziergangswissenschaft* (pp. 49–56). Kassel: Martin Schmitz Verlag.

Burckhardt, L. (2006). *Warum ist Landschaft schön? Die Spaziergangswissenschaft*. Kassel: Martin Schmitz Verlag.

Burnet, J. (1962). *Platonis Opera recognovit brevique adnotatione critica instruxit Ioannes Burnet. Tomus IV*. Oxonii: Clarendon.

Buttlar, A. v. (1989). *Der Landschaftsgarten. Gartenkunst des Klassizismus und der Romantik*. Köln: DuMont.

Büttner, N. (2006). *Geschichte der Landschaftsmalerei*. München: Hirmer.

Büttner, N. (2019). Landschaftsmalerei. In O. Kühne, F. Weber, K. Berr, & C. Jenal (Eds.), *Handbuch Landschaft* (pp. 577–584). Wiesbaden: Springer VS.

Callon, M. (1999). Actor-network theory – The market test. *The Sociological Review, 47*(S1), 181–195.

Carus, C. G. (1982). *Briefe und Aufsätze über Landschaftsmalerei*. Leipzig: Kiepenheuer. (Hrsg. und mit einem Nachwort von Gertrud Heider).

Chilla, T., Kühne, O., Weber, F., & Weber, F. (2015). „Neopragmatische" Argumente zur Vereinbarkeit von konzeptioneller Diskussion und Praxis der Regionalentwicklung. In O. Kühne & F. Weber (Eds.), *Bausteine der Regionalentwicklung* (pp. 13–24). Wiesbaden: Springer VS.

Chilla, T., Kühne, O., & Neufeld, M. (2016). *Regionalentwicklung*. Stuttgart: Ulmer.

Conwentz, H. (1904). *Die Gefährdung der Naturdenkmäler und Vorschläge zu ihrer Erhaltung. Denkschrift*. Berlin: Gebrüder Borntraeger.

Conzen, M. P. (Ed.). (2010). *The making of the American landscape* (2nd ed.). New York: Routledge.

Cosgrove, D. (1985). Prospect, perspective and the evolution of the landscape idea. *Transactions of the Institute of British Geographers, 10*(1), 45–62. https://doi.org/10.2307/622249.

Cosgrove, D. (1989). A terrain of metaphor: Cultural geography 1988–89. *Progress in Human Geography, 13*(4), 566–575. https://doi.org/10.1177/030913258901300406.

Cosgrove, D. (2004). Landscape and Landschaft. *German Historical Institut Bulletin, 35*(Fall), 57–71.

Cosgrove, D., & Daniels, S. (Eds.). (1988). *The iconography of landscape. Essays on the symbolic representation, design and use of past environments* (Cambridge studies in historical geography, Vol. 9). Cambridge: Cambridge University Press.

Cosgrove, D., & Jackson, P. (1987). New directions in cultural geography. *Area, 19*(2), 95–101.

Cosgrove, D. E. (1984). *Social formation and symbolic landscape*. London: University of Wisconsin Press.

Cosgrove, D. E. (1993). *The palladian landscape. Geographical change and its cultural representations in sixteenth-century italy*. University Park: Pennsylvania State University Press.

Cosgrove, D. E. (1998). *Social formation and symbolic landscape*. Wisconsin: University of Wisconsin Press.

Cronon, W. (1996). Introduction: In search of nature. In W. Cronon (Ed.), *Uncommon ground. Rethinking the human place in nature* (pp. 23–68). New York: W. W. Norton & Company.

Curtius, E. R. (1954 [1948]). *Europäische Literatur und lateinisches Mittelalter* (Zweite, durchgesehene Aufl.). Bern: Francke.

Dahrendorf, R. (1972). *Konflikt und Freiheit. Auf dem Weg zur Dienstklassengesellschaft*. München: Piper.

Daniels, S. (1988). The political iconography of woodland in later Georgian England. In D. Cosgrove & S. Daniels (Eds.), *The iconography of landscape. Essays on the symbolic representation, design and use of past environments* (Cambridge studies in historical geography, Vol. 9, pp. 43–82). Cambridge: Cambridge University Press.

Daniels, S. (1999). *Humphry repton. Landscape gardening and the geography of georgian England.* New Haven: Yale University Press.

Defila, R., & Di Giulio, A. (1998). Interdisziplinarität und Disziplinarität. In J.-H. Olbertz (Ed.), *Zwischen den Fächern – über den Dingen? Universalisierung versus Spezialisierung akademischer Bildung* (pp. 111–137). Wiesbaden: Leske + Budrich.

Denzer, V. (2019). Landschaft als Text. In O. Kühne, F. Weber, K. Berr, & C. Jenal (Eds.), *Handbuch Landschaft* (pp. 81–89). Wiesbaden: Springer VS.

Dettmar, J. (2018). Wissenschaftliche Grundlagen der Landschaftsarchitektur. In K. Berr (Ed.), *Landschaftsarchitekturtheorie. Aktuelle Zugänge, Perspektiven und Positionen* (pp. 21–50). Wiesbaden: Springer VS.

Diaz-Bone, R. (2010). *Kulturwelt, Diskurs und Lebensstil. Eine diskurstheoretische Erweiterung der Bourdieuschen Distinktionstheorie* (2., erweiterte ed.). Wiesbaden: VS Verlag.

Dieckmann, W. (1975). *Sprache in der Politik. Einführung in die Pragmatik und Semantik der politischen Sprache* (Sprachwissenschaftliche Studienbücher, 2. Aufl.). Heidelberg: Winter (Mit einem Literaturbericht zur 2. Aufl.)

Dilthey, W. (1914). *Gesammelte Schriften, Band 2. Weltanschauung und Analyse des Menschen seit Renaissance und Reformation.* Leipzig: B.G. Teubner.

Dinnebier, A. (1997). Der Barockgarten als Vorbereitung der Idee der „Landschaft". *Stadt+Grün, 46,* 157–164.

Doktor, W. (1975). *Die Kritik der Empfindsamkeit* (Regensburger Beiträge zur deutschen Sprach- und Literaturwissenschaft Reihe B, Untersuchungen 5). Bern: Lang.

Donnelly, K. J. (2002). A ramble through the margins of the cityscape: The postmodern as the return of nature. In M. J. Dear & S. Flusty (Eds.), *The spaces of postmodernity. Readings in human geography* (pp. 423–430). Oxford: Blackwell Publishers.

Doucet, I., & Janssens, N. (2011). *Transdisciplinary knowledge production in architecture and urbanism. Towards hybrid modes of inquiry* (Urban and landscape perspectives, Vol. 11). Dordrecht: Springer Science+Business Media B.V.

Drexler, D. (2009a). Kulturelle Differenzen der Landschaftswahrnehmung in England, Frankreich, Deutschland und Ungarn. In T. Kirchhoff & L. Trepl (Eds.), *Vieldeutige Natur. Landschaft, Wildnis und Ökosystem als kulturgeschichtliche Phänomene* (Sozialtheorie, pp. 119–136). Bielefeld: transcript.

Drexler, D. (2009b). *Landschaft und Landschaftswahrnehmung: Untersuchung des kulturhistorischen Bedeutungswandels von Landschaft anhand eines Vergleichs von England, Frankreich, Deutschland und Ungarn.* (Dissertation, Technische Universität München). München. https://mediatum.ub.tum.de/doc/738822/738822.pdf. Accessed 16 Mar 2017.

Drexler, D. (2010). *Landschaften und Landschaftswahrnehmung: Untersuchung des kulturhistorischen Bedeutungswandels von Landschaft anhand eines Vergleichs von England, Frankreich, Deutschland und Ungarn.* München: Technische Universität München.

Drexler, D. (2013). Die Wahrnehmung der Landschaft – ein Blick auf das englische, französische und ungarische Landschaftsverständnis. In D. Bruns & O. Kühne (Eds.), *Landschaften: Theorie, Praxis und internationale Bezüge* (pp. 37–54). Schwerin: Oceano Verlag.

Duncan, J., & Duncan, N. (1988). (Re)reading the landscape. *Environment and Planning D: Society and Space, 6*(2), 117–126. https://doi.org/10.1068/d060117.

Duncan, N., & Duncan, J. (2009). Doing landscape interpretation. In D. DeLyser, S. Herbert, S. Aitken, M. A. Crang, & L. McDowell (Eds.), *The SAGE handbook of qualitative geography.*

London: SAGE Publications. http://www.sage-ereference.com/hdbk_qualgeography/Article_n13.html. Accessed 7 Jan 2019.

Duncan, J. S. (1980). The superorganic in American cultural geography. *Annals of the Association of American Geographers, 70*(2), 181–198.

Duncan, J. S. (1990). *The city as text: The politics of landscape interpretation in the kandyan kingdom.* Cambridge: Cambridge University Press.

Dutt, C. (2011). Historische Semantik als Begriffsgeschichte. Theoretische Grundlagen und paradigmatische Anwendungsfelder. In J. Riecke (Ed.), *Historische Semantik* (pp. 37–50). Berlin: de Gruyter.

Eagleton, T. (1994). *Ästhetik. Die Geschichte ihrer Ideologie.* Stuttgart: J B. Metzler.

Eberle, M. (1980). *Individuum und Landschaft. Zur Entstehung und Entwicklung der Landschaftsmalerei.* Gießen: Anabas-Verlag.

Eckardt, F. (2014). *Stadtforschung. Gegenstand und Methoden.* Wiesbaden: Springer VS.

Eisel, U. (1980). *Die Entwicklung der Anthropogeographie von einer „Raumwissenschaft" zur Gesellschaftswissenschaft.* Kassel: Gesamthochschul-Bibliothek.

Eisel, U. (1982). Die schöne Landschaft als kritische Utopie oder als konservatives Relikt. Über die Kristallisation gegnerischer politischer Philosophien im Symbol „Landschaft". *Soziale Welt, 33*(2), 157–168.

Eisel, U. (1992a). Über den Umgang mit dem Unmöglichen. Ein Erfahrungsbericht über interdisziplinäre Studienprojekte in der Landschaftsplanung. Teil 1 und 2. http://ueisel.de/fileadmin/dokumente/ausgetauscht%20ab%20november%202009/Ueber_den_Umgang_mit_dem_Unmoeglichen_INTERDIS_1992. Accessed 18 Jan 2019.

Eisel, U. (1992b). Über den Umgang mit dem Unmöglichen. Ein Erfahrungsbericht über Interdisziplinarität im Studiengang Landschaftsplanung – Teil 2. *Gartenamt, 41*(10), 710–719.

Eisel, U. (1992c). Über den Umgang mit dem Unmöglichen. Ein Erfahrungsbericht über Interdisziplinarität im Studiengang Landschaftsplanung – Teil 1. *Gartenamt, 41*(9), 593–605.

Eisel, U. (1997). Unbestimmte Stimmungen und bestimmte Unstimmigkeiten. Über die guten Gründe der deutschen Landschaftsarchitektur für die Abwendung von der Wissenschaft und die schlechten Gründe für ihre intellektuelle Abstinenz – mit Folgerungen für die Ausbildung in diesem Fach. In S. Bernhard, & P. Sattler (Eds.), *Vor der Tür. Aktuelle Landschaftsarchitekru aus Berlin* (pp. 17–33). München: Callwey. http://www.ueisel.de/fileadmin/dokumente/eisel/Unbestimmte_Stimmungen/Eisel_Unbestimmte_Stimmungen_fertig.pdf. Accessed 12 January 2019.

Eisel, U. (2009). *Landschaft und Gesellschaft. Räumliches Denken im Visier* (Raumproduktionen: Theorie und gesellschaftliche Praxis, Bd. 5). Münster: Westfälisches Dampfboot.

Eisel, U., & Körner, S. (Eds.). (2009). *Befreite Landschaft. Moderne Landschaftsarchitektur ohne arkadischen Ballast?* (Beiträge zur Kulturgeschichte der Natur, Vol. 18). Freising: Technische Universität München.

Elliger, W. (1975). *Die Darstellung der Landschaft in der griechischen Dichtung.* Berlin: de Gruyter.

Erb, G. (1997). *Die Landschaftsdarstellung in der deutschen Druckgraphik vor Albrecht Dürer* (Europäische Hochschulschriften Reihe 28, Kunstgeschichte, Vol. 282). Frankfurt a. M.: Lang.

Eschenbach, W. v. (1998 [first around 1200/1210]). *Parzival* (Studienausgabe). Berlin: de Gruyter.

Falter, R. (1999). Natur als Landschaft und als Gott. Fluß- und Berggötter in der Spätantike. In R. P. Sieferle & H. Breuninger (Eds.), *Natur-Bilder. Wahrnehmungen von Natur und Umwelt in der Geschichte* (pp. 137–179). Frankfurt a. M.: Campus.

Fehn, K. (1976). Historische Geographie. Eigenständige Wissenschaft und Teilwissenschaft der Geographie. *Mitteilungen der Geogrphischen Gesellschaft München, 61,* 35–51.

Felten, F. J., Müller, H., & Ochs, H. (Eds.). (2012). *Landschaft(en). Begriffe – Formen – Implikationen* (Geschichtliche Landeskunde, Vol. 68). Stuttgart:Steiner.

Fischer, C., Kuhn, H., & Bernt, G. (1974). *Carmina Burana. Die Lieder der Benediktbeurer Handschrift* (Ausgabe des Originaltextes nach der von Bernhard Bischoff abgeschlossenen kristischen Ausgabe von Alfons Hilka und Otto Schuhmann). Zürich: Artemis.

Fischer, L. (2001). Das Erhabene und die ,feinen Unterschiede'. Zur Dialektik in den soziokulturellen Funktionen von ästhetischen Deutungen der Landschaft. In R. W. Brednich, A. Schneider, & U. Werner (Eds.), *Natur – Kultur. Volkskundliche Perspektiven auf Mensch und Umwelt* (pp. 347–356). Münster: Waxmann.

Fischer, L. (Ed.). (2004). *Projektionsfläche Natur. Zum Zusammenhang von Naturbildern und gesellschaftlichen Verhältnissen*. Hamburg: Hamburg University Press.

Fischer, L. (2007). Kulturlandschaft – Naturtheoretische und kultursoziologische Anmerkungen zu einem Konzept. In Stiftung Natur und Umwelt Rheinland-Pfalz (Ed.), *Landschaftskult - Kulturlandschaft* (Denkanstöße, Vol. 6, pp. 16–27). Mainz: Stiftung Natur und Umwelt Rheinland-Pfalz.

Flasch, K. (1980). *Augustin. Einführung in sein Denken* (Reclams Universalbibliothek, Vol. 9962). Stuttgart: Reclam.

Fontaine, D. (2017). *Simulierte Landschaften in der Postmoderne. Reflexionen und Befunde zu Disneyland, Wolfersheim und GTA V*. Wiesbaden: Springer VS.

Foucault, M. (1977). *Überwachen und Strafen. Die Geburt des Gefängnisses* (Suhrkamp-Taschenbuch Wissenschaft, Vol. 184). Frankfurt a. M.: Suhrkamp.

Foxley, A. (2010). *Distance & engagement. Walking, thinking and making landscape*. Baden: Lars Müller Publishers.

Franck, G. (2007). *Ökonomie der Aufmerksamkeit. Ein Entwurf*. München: dtv.

Frank, S., Fürst, C., Koschke, L., Witt, A., & Makeschin, F. (2013). Assessment of landscape aesthetics – Validation of a landscape metrics-based assessment by visual estimation of the scenic beauty. *Ecological Indicators, 32*(9), 222–231.

Franzen, B. (2003). Landscape Drift – Thesen zu einer zeitgenössischen Kultur der Landschaft. In I. Flagge (Ed.), *Architektur und Wahrnehmung. Jahrbuch Licht und Architektur 2003* (pp. 122–125). Darmstadt: Das Beispiel.

Franzen, B., & Krebs, S. (Eds.). (2005). *Landschaftstheorie Texte der Cultural Landscape Studies* (Kunstwissenschaftliche Bibliothek, Vol. 26). Köln: König.

Freeman, R. E. (1984). *Strategic management. A stakeholder approach*. Boston: Pitman.

Freyer, H. (1990). Landschaft und Geschichte. In G. Gröning & U. Herlyn (Eds.), *Landschaftswahrnehmung und Landschaftserfahrung. Texte zur Konstitution und Rezeption von Natur als Landschaft* (pp. 43–65). München: Minerva.

Friedländer, L. (1867). Darstellung aus der Sittengeschichte Roms (Zweite vermehrte Aufl., Zweiter Teil). Leipzig: Hirzel.

Friedländer, M. J. (1947). Die Sonderart des Landschaftlichen. In M. J. Friedländer (Ed.), *Eassays über die Landschaftsmalerei und andere Bildgattungen* (pp. 9–23). Den Haar: A. A. M. Stols.

Frizell, B. S. (2009). *Arkadien. Mythos und Wirklichkeit* (Aus dem Schwedischen übersetzt von Ylva Eriksson-Kuchenbuch). Köln: Böhlau.

Fuhrer, U., & Wölfing, S. (1997). *Von den sozialen Grundlagen des Umweltbewußtseins zum verantwortlichen Umwelthandeln. Die sozialpsychologische Dimension globaler Umweltproblematik*. Bern: Huber.

Fürst, D., Gailing, L., Pollermann, K., & Röhring, A. (Eds.). (2008). *Kulturlandschaft als Handlungsraum. Institutionen und Governance im Umgang mit dem regionalen Gemeinschaftsgut Kulturlandschaft*. Dortmund: Verlag Dorothea Rohn.

Gabriel, G. (2005). Orientierung – Unterscheidung – Vergegenwärtigung. Zur Unverzichtbarkeit nicht propositionaler Erenntnis für die Philosophie. In G. Wolters & M. Carrier (Eds.), *Homo

Sapiens und Homo Faber. Epistemische und technische Rationalität in Antike und Gegenwart. Festschrift für Jürgen Mittelstraß (pp. 323–334). Berlin: De Gruyter.

Gailing, L. (2012). Sektorale Institutionensysteme und die Governance kulturlandschaftlicher Handlungsräume. Eine institutionen- und steuerungstheoretische Perspektive auf die Konstruktion von Kulturlandschaft. *Raumforschung und Raumordnung, 70*(2), 147–160. https://doi.org/10.1007/s13147-011-0135-x.

Gailing, L. (2015). Energiewende als Mehrebenen-Governance. *Nachrichten der ARL, 45*(2), 7–10.

Gailing, L. (2018). Die räumliche Governance der Energiewende: Eine Systematisierung der relevanten Governance-Formen. In O. Kühne & F. Weber (Eds.), *Bausteine der Energiewende* (pp. 75–90). Wiesbaden: Springer VS.

Gailing, L., Keim, K.-D., & Röhring, A. (2006). *Analyse von informellen und dezentralen Institutionen und Public Governance mit kulturlandschaftlichem Hintergrund in der Beispielregion Barnim* (Materialien no. 6), Berlin: Berlin-Brandenburgische Akademie der Wissenschaften.

Gebhard, G., Geisler, G., & Schröter, S. (2007). *Heimat Konturen und Konjunkturen eines umstrittenen Konzepts*. Bielefeld: transcript.

Gebhardt, H. (2016). Entwicklungspfade und Perspektiven der Humangeographie im deutschsprachigen Raum – Einige Leitlinien. In J. Aistleitner, M. Coy, & J. Stötter (Eds.), *Die Welt verstehen – Eine geographische Herausforderung. Eine Festschrift der Geographie Innsbruck für Axel Borsdorf* (Innsbrucker geographische Studien, Vol. 40, pp. 43–59). Innsbruck: Geographie Innsbruck Selbstverlag.

Gebhardt, H. (2019). Landeskunde und Landschaft – Eine kritische Betrachtung. In O. Kühne, F. Weber, K. Berr, & C. Jenal (Eds.), *Handbuch Landschaft* (pp. 289–298). Wiesbaden: Springer VS.

Gebser, J. (1966). *Ursprung und Gegenwart, Fundamente und Manifestationen der aperspektivischen Welt*. Stuttgart: Deutsche Verlags-Anstalt.

Gehring, K., & Kohsaka, R. (2007). ‚Landscape' in the Japanese language: Conceptual differences and implications for landscape research. *Landscape Research, 32*(2), 273–283. https://doi.org/10.1080/01426390701231887.

Gelfert, H.-D. (2000). *Was ist Kitsch?* (Kleine Reihe V und R, Vol. 4024). Göttingen: Vandenhoeck und Ruprecht.

Gethmann, C. F. (1991). Vielheit der Wissenschaften – Einheit der Lebenswelt. In Akademie der Wissenschaften zu Berlin (Ed.), *Einheit der Wissenschaften* (pp. 349–371). Berlin: De Gruyter.

Gethmann, C. F. (2005). Ist das Wahre das Ganze? Methodologische Probleme Integrierter Forschung. In G. Wolters & M. Carrier (Eds.), *Homo Sapiens und Homo Faber. Epistemische und technische Rationalität in Antike und Gegenwart. Festschrift für Jürgen Mittelstraß* (pp. 391–404). Berlin: De Gruyter.

Gethmann, C. F. (2010). Die Aktualität Methodischen Denkens. In C. F. Gethmann, & J. Mittelstraß (Eds.), *Paul Lorenzen zu Ehren* (Konstanzer Universitätsreden, Vol. 241, pp. 15–37). Konstanz: Universitätsverlag Konstanz.

Gethmann, C. F. (2011). Philosophie – Zwischen Lebenswelt und Wissenschaft. In C. F. Gethmann, J. C. Bottek, & S. Hiekel (Eds.), *Lebenswelt und Wissenschaft. XXI. Deutscher Kongreß für Philosophie 15.–19. September 2008 an der Universität Duisburg – Essen*. Kolloquienbeiträge (pp. 3–16). Hamburg: Felix Meiner Verlag.

Gethmann-Siefert, A. (1995). *Einführung in die Ästhetik* (UTB, Vol. 1875). München: Fink.

Gethmann-Siefert, A. (2005). *Einführung in Hegels Ästhetik*. München: Fink.

Gieryn, T. F. (1983). Boundary-work and the demarcation of science from non-science: Strains and interests in professional ideologies of scientists. *American Sociological Review, 48*(6), 781–795. https://doi.org/10.2307/2095325.

Gießmann, S., & Taha, N. (Eds.). (2017). *Susan Leigh Star. Grenzobjekte und Medienforschung* (Locating Media/Situierte Medien, Vol. 10). Bielefeld: transcript.

Gil, T. (2000). *Der Begriff der ästhetischen Erfahrung*. Berlin: Berlin Verlag.

Glasersfeld, E. v. (2000). Problems of constructivism. In L. P. Steffe, & P. W. Thompson (Eds.), *Radical constructivism in action. Building on the pioneering work of Ernst von Glasersfeld* (Studies in mathematics education series, Vol. 15, pp. 3–9). New York: Falmer.

Glasze, G. (2015). Identitäten und Räume als politisch: Die Perspektive der Diskurs- und Hegemonietheorie. *Europa regional, 21*(2013), 1–2–23–34.

Glasze, G., & Mattissek, A. (2009). Diskursforschung in der Humangeographie: Konzeptionelle Grundlagen und empirische Operationalisierung. In G. Glasze & A. Mattissek (Eds.), *Handbuch Diskurs und Raum. Theorien und Methoden für die Humangeographie sowie die sozial- und kulturwissenschaftliche Raumforschung* (pp. 11–59). Bielefeld: transcript.

Goethe, J. W. v. (1985). *Die Leiden des jungen Werther*. Stuttgart: Reclam (herausgegeben von Kurt Rothmann).

Goethe, J. W. v. (1999). *Poetische Werke. Band 4. Frühe Dramen, Bruchstücke, Übersetzungen und Bearbeitungen*. Essen: Phaidon Verlag.

Goethe, J. W. v., Schiller, F., & Meyer, H. (1896). Über den Dilettantismus. In Großherzogin Sophie von Sachsen (Ed.), *Goethes Werke* (Vol. 47, pp. 299–326). Weimar: Böhlau.

Grassi, E. (1962). *Die Theorie des Schönen in der Antike* (Geschichte der Ästhetik, Vol. 1). Köln: DuMont.

Grassi, E. (1980). *Die Theorie des Schönen in der Antike* (DuMont Taschenbücher, Vol. 90, Überarbeitete Neuausgabe). Köln: DuMont.

Greider, T., & Garkovich, L. (1994). Landscapes: The social construction of nature and the environment. *Rural Sociology, 59*(1), 1–24. https://doi.org/10.1111/j.1549-0831.1994.tb00519.x.

Greverus, I.-M. (1979). *Auf der Suche nach Heimat* (Beck'sche Schwarze Reihe, Vol. 189). München: Beck.

Groh, R., & Groh, D. (1992). Petrarca und der Mont Ventoux. *Merkur, 46*(4), 290–307.

Groh, R., & Groh, D. (1996). *Die Außenwelt der Innenwelt. Zur Kulturgeschichte der Natur 2*. Frankfurt a. M.: Suhrkamp.

Großheim, M. (1999). Atmosphären in der Natur: Phänomene oder Konstrukte? In R. P. Sieferle & H. Breuninger (Eds.), *Natur-Bilder. Wahrnehmungen von Natur und Umwelt in der Geschichte* (pp. 325–365). Frankfurt a. M.: Campus.

Groth, P., & Wilson, C. (2005). Die Polyphonie der Cultural Landscape Studies [2003]. In B. Franzen, & S. Krebs (Eds.), *Landschaftstheorie. Texte der Cultural Landscape Studies* (Kunstwissenschaftliche Bibliothek, Vol. 26, pp. 58–90). Köln: König.

Gruenter, R. (1975a [1953]). Landschaft. Bemerkungen zu Wort und Bedeutungsgeschichte. In A. Ritter (Ed.), *Landschaft und Raum in der Erzählkunst* (Wege der Forschung, Vol. 418, pp. 192–207). Darmstadt: WBG.

Gruenter, R. (1975b [1953]). Zum Problem der Landschaftsdarstellung im höfischen Versroman. In A. Ritter (Ed.), *Landschaft und Raum in der Erzählkunst* (Wege der Forschung, Vol. 418, pp. 193–335). Darmstadt: WBG.

Grunwald, A., & Schmidt, J. C. (2005). Method(olog)ische Fragen der Inter- und Transdisziplinarität. Wege zu einer praxisstützenden Interdisziplinaritätsforschung. *Technikfolgenabschätzung – Theorie und Praxis, 14*(2), 4–11.

Gumbrecht, H. U. (2006). *Dimensionen und Grenzen der Begriffsgeschichte*. München: Wilhelm Fink.

Gutmann, M. (2005). Disziplinarität und Inter-Diziplinarität in methodologischer Sicht. *Technikfolgenabschätzung – Theorie und Praxis, 14*, 69–73.

Haber, W. (1993). Über die Entwicklung der Naturschutzgesetzgebung. In Bayerische Akademie der Wissenschaften (Ed.), *Probleme der Umweltforschung in historischer Sicht. Rundgespräch am 16. und 17. November 1992 in München* (Rundgespräche der Kommission für Ökologie, Vol. 7, pp. 221–231). München: Dr. Friedrich Pfeil.

Haber, W. (2005). Pflege des Landes – Verantwortung für Landschaft und Heimat. In Deutscher Rat für Landespflege (Ed.), *Landschaft und Heimat* (Schriftenreihe des Deutschen Rates für Landespflege, Heft 77, pp. 100–107). Meckenheim: DCM.

Haber, W. (2006). Landschaft zwischen Bild und Ressource. In H. Gaese, S. Sandholz, & A. Böhler (Eds.), *Denken in Räumen. Nachhaltiges Ressourcenmanagement als Identitätssicherung – Durch Veränderung der Rahmenbedingungen gefährdete Kulturlandschaften und das Problem ihrer Erhaltung* (pp. 275–288). Stohlberg-Venwegen: Zillekens.

Haber, W. (2007). Vorstellungen über Landschaft. In B. Busch (Ed.), *Jetzt ist die Landschaft ein Katalog voller Wörter. Beiträge zur Sprache der Ökologie* (Valerio, Vol. 5, pp. 78–85). Göttingen: Wallstein.

Hage, G., & Bäumer, C. (2019). Landschaftsplanung. In O. Kühne, F. Weber, K. Berr, & C. Jenal (Eds.), *Handbuch Landschaft* (pp. 245–264). Wiesbaden: Springer VS.

Hahn, A. (Ed.). (2012). *Erlebnislandschaft – Erlebnis Landschaft? Atmosphären im architektonischen Entwurf* (Architekturen, Vol. 13). Bielefeld: transcript.

Hard, G. (1969). Das Wort Landschaft und sein semantischer Hof. Zur Methode und Ergebnis eines linguistischen Tests. *Wirkendes Wort, 19,* 3–14.

Hard, G. (1970a). Der 'Totalcharakter der Landschaft'. Re-Interpretation einiger Textstellen bei Alexander von Humboldt. *Erdkundliches Wissen, Beiheft,* 49–71.

Hard, G. (1970b). *Die „Landschaft" der Sprache und die „Landschaft" der Geographen. Semantische und forschungslogische Studien.* Bonn: Ferdinand Dümmlers Verlag.

Hard, G. (1970c). „Was ist eine Landschaft?". Über Etymologie als Denkform in der geographischen Literatur. In D. Bartels (Ed.), *Wirtschafts- und Sozialgeographie* (Neue wissenschaftliche Bibliothek, Vol. 35, pp. 66–84). Köln: Kiepenheuer & Witsch.

Hard, G. (1973). *Die Geographie. Eine wissenschaftstheoretische Einführung* (Sammlung Göschen). Berlin: De Gruyter.

Hard, G. (1977). Zu den Landschaftsbegriffen der Geographie. In A. Hartlieb von Wallthor, & H. Quirin (Eds.), *„Landschaft" als interdisziplinäres Forschungsproblem. Vorträge und Diskussionen des Kolloquiums am 7./8. November 1975 in Münster* (pp. 13–24). Münster: Aschendorff.

Hard, G. (1991). Landschaft als professionelles Idol. *Garten + Landschaft, 3,* 13–18.

Hard, G. (2002a). Die „Natur" der Geographen. In U. Luig, & H.-D. Schultz (Eds.), *Natur in der Moderne. Interdisziplinäre Ansichten* (Berliner geographische Arbeiten, Vol. 93, pp. 67–86). Berlin: Geographisches Institut der Humboldt-Universität.

Hard, G. (Ed.). (2002b). *Landschaft und Raum. Aufsätze zur Theorie der Geographie* (Osnabrücker Studien zur Geographie, Vol. 22). Osnabrück: Universitätsverlag Rasch.

Hard, G. (2002c [1983]). Zu Begriff und Geschichte von „Natur" und „Landschaft" in der Geographie des 19. und 20. Jahrhunderts. In G. Hard (Ed.), *Landschaft und Raum. Aufsätze zur Theorie der Geographie* (Osnabrücker Studien zur Geographie, Vol. 22, pp. 171–210). Osnabrück: Universitätsverlag Rasch.

Hard, G. (2003). Studium in einer diffusen Disziplin. In G. Hard (Ed.), *Dimensionen geographischen Denkens. Aufsätze zur Theorie der Geographie* (Osnabrücker Studien zur Geographie, Vol. 23, pp. 173–230). Göttingen: V & R Unipress.

Hard, G., & Gliedner, A. (1977). Wort und Begriff Landschaft anno 1976. In F. Achleitner (Ed.), *Die Ware Landschaft. Eine kritische Analyse des Landschaftsbegriffs* (pp. 16–24). Salzburg: Residenz Verlag.

Hartke, W. (1956). Die „Sozialbrache" als Phänomen der geographischen Differenzierung der Landschaft. *Erdkunde, 10*(4), 257–269.

Hartlieb von Wallthor, A., & Quirin, H. (Eds.). (1977). *„Landschaft" als interdisziplinäres Forschungsproblem. Vorträge und Diskussionen des Kolloquiums am 7./8. November 1975 in Münster*. Münster: Aschendorff.

Hartz, A. M. (2019). Landschaft als Belang der Regionalplanung. In O. Kühne, F. Weber, K. Berr, & C. Jenal (Eds.), *Handbuch Landschaft* (pp. 265–278). Wiesbaden: Springer VS.

Harvey, D. (1996). *Justice, nature and the geography of difference*. Malden: Blackwell.

Harvey, D. C. (2001). Heritage pasts and heritage presents: Temporality, meaning and the scope of heritage studies. *International Journal of Heritage Studies, 7*(4), 319–338. https://doi.org/10.1080/13581650120105534.

Harvey, D., Wilkinson, & Timothy J. (2018). Landscape and heritage: emerging landscapes of heritage. In P. Howard, I. H. Thompson, E. Waterton, & M. Atha (Eds.), *The routledge companion to landscape studies* (Routledge companions, 2nd ed., pp. 176–191). Abingdon: Routledge.

Hasse, J. (1993). *Heimat und Landschaft. Über Gartenzwerge, Center Parcs und andere Ästhetisierungen*. Wien: Passagen-Verlag.

Hauck, T. E. (2014). *Landschaft und Gestaltung. Die Vergegenständlichung ästhetischer Ideen am Beispiel von „Landschaft"*. Bielefeld: transcript.

Hauser, S., & Kamleithner, C. (2006). *Ästhetik der Agglomeration* (Zwischenstadt, Vol. 8). Wuppertal: Müller + Busmann.

Hauser, S. & Weber, J. (2015). *Architektur in transdisziplinärer Perspektive. Von Philosophie bis Tanz Aktuelle Zugänge und Positionen* (Architekturen, Vol. 23). Bielefeld: transcript.

Hauskeller, M. (Ed.). (1995). *Was das Schöne sei. Klassische Texte von Platon bis Adorno*. München: Deutscher Taschenbuch Verlag.

Hauskeller, M. (2005). *Was ist Kunst? Positionen der Ästhetik von Platon bis Danto* (Beck'sche Reihe, 8th ed.). München: Beck.

Heckhausen, H. (1987). ‚Interdisziplinäre Forschung' zwischen Intra-, Multi- und Chimären-Disziplinarität. In J. Kocka (Ed.), *Interdisziplinarität. Praxis, Herausforderung, Ideologie* (pp. 129–145). Frankfurt a. M.: Suhrkamp.

Hegel, G. W. F. (1980). *Phänomenologie des Geistes* (Gesammelte Werke, Vol. 9). Hamburg: Meiner.

Hegel, G. W. F. (1989). Text 35: Bericht über eine Alpenwanderung. In Rheinisch-Westfälische Akademie der Wissenschaften (Ed.), *Georg Wilhelm Friedrich Hegel. Gesammelte Werke. Band 1* (pp. 381–398). Hamburg: Meiner.

Hegel, G. W. F. (1995). *Grundlinien der Philosophie des Rechts* (Philosophische Bibliothek, Band 483, 5., neu durchgesehene Auflage). Hamburg: Meiner.

Hegel, G. W. F. (2003 [1823]). *Vorlesungen über die Philosophie der Kunst. Berlin 1823*. Hamburg: Meiner.

Heidegger, M. (1963 [1938]). *Holzwege* (4th ed.). Frankfurt a. M.: Klostermann.

Heidegger, M. (1993 [1927]). *Sein und Zeit*. Tübingen: Max Niemeyer Verlag.

Heiland, S. (2010). Kulturlandschaft. In D. Henckel, K. von Kuczkowski, P. Lau, E. Pahl-Weber, & F. Stellmacher (Eds.), *Planen – Bauen – Umwelt. Ein Handbuch* (pp. 278–283). Wiesbaden: VS Verlag für Sozialwissenschaften.

Heiland, S. (2019). Kulturlandschaft. In O. Kühne, F. Weber, K. Berr, & C. Jenal (Eds.), *Handbuch Landschaft* (pp. 651–665). Wiesbaden: Springer VS.

Held, K. (1991). Husserls neue Einführung in die Philosophie: der Begriff der Lebenswelt. In C. F. Gethmann (Ed.), *Lebenswelt und Wissenschaft. Studien zum Verhältnis von Phänomenologie und Wissenschaftstheorie* (Neuzeit und Gegenwart, Vol. 1, pp. 79–113). Bonn: Bouvier.

Hengst, L. (2019). Iconoscape – Bestimmungen eines neuen Forschungsbegriffs für neue Formationen von Landschaftsbildlichkeit. In O. Kühne, F. Weber, K. Berr, & C. Jenal (Eds.), *Handbuch Landschaft* (pp. 461–468). Wiesbaden: Springer VS.

Hennebo, D. (1987). *Gärten des Mittelalters*. München: Artemis.

Hennecke, S. (2019). Freiraumplanung. In O. Kühne, F. Weber, K. Berr, & C. Jenal (Eds.), *Handbuch Landschaft* (pp. 219–229). Wiesbaden: Springer VS.

Herder, J. G. (1964). *Ideen zur Philosophie der Geschichte der Menschheit*. Berlin: Aufbau-Verlag (Herders Werke in fünf Bänden. Vierter Band).

Hernik, J., & Dixon-Gough, R. (2013). The concept and importance of landscape in Polish language and in Poland. In D. Bruns & O. Kühne (Eds.), *Landschaften: Theorie, Praxis und internationale Bezüge* (pp. 83–98). Schwerin: Oceano Verlag.

Hinrichs, W. (1974). Heimat, Heimatkunde. In J. Ritter, K. Gründer, & G. Gabriel (Eds.), *Historisches Wörterbuch der Philosophie. Band 3 G–H* (pp. 1037–1039). Basel: Schwabe.

Hirsch, E. (1995). Hortus Oeconomicus: Nutzen, Schönheit, Bildung. Das Dessau-Wörlitzer Gartenreich als Landschaftsgestaltung der euopäischen Aufklärung. In H. Wunderlich (Ed.), *„Landschaft" und Landschaften im achtzehnten Jahrhundert. Tagung der Deutschen Gesellschaft für die Erforschung des 18. Jahrhunderts, Herzog-August-Bibliothek Wolfenbüttel, 20. bis 23. November 1991* (Beiträge zur Geschichte der Literatur und Kunst des 18. Jahrhunderts, Vol. 13, pp. 179–208). Heidelberg: Universitätsverlag C. Winter.

Hoeft, C., Messinger-Zimmer, S., & Zilles, J. (Eds.). (2017). *Bürgerproteste in Zeiten der Energiewende. Lokale Konflikte um Windkraft, Stromtrassen und Fracking*. Bielefeld: transcript.

Hoeres, W. (2004). *Der Weg der Anschauung. Landschaft zwischen Ästhetik und Metaphysik* (Die Graue Reihe, Vol. 40). Dietzenbach: Die Graue Edition.

Hoffmeister, J. (Ed.). (1969). *Briefe von und an Hegel. Band II: 1813–1822*. Hamburg: Meiner.

Hofmeister, S., & Mölders, T. (2019). StadtLandschaft. In O. Kühne, F. Weber, K. Berr, & C. Jenal (Eds.), *Handbuch Landschaft* (pp. 731–741). Wiesbaden: Springer VS.

Hohl, H. (1977). Das Thema Landschaft in der deutschen Malerei des ausgehenden 18. und beginnenden 19. Jahrhunderts. In A. Hartlieb von Wallthor, & H. Quirin (Eds.), *„Landschaft" als interdisziplinäres Forschungsproblem. Vorträge und Diskussionen des Kolloquiums am 7./8. November 1975 in Münster* (pp. 45–53). Münster: Aschendorff.

Hokema, D. (2009). Die Landschaft der Regionalentwicklung: Wie flexibel ist der Landschaftsbegriff? *Raumforschung und Raumordnung, 67*(3), 239–249.

Hokema, D. (2013). *Landschaft im Wandel? Zeitgenössische Landschaftsbegriffe in Wissenschaft, Planung und Alltag*. Wiesbaden: Springer VS.

Hokema, D. (2015). Landscape is everywhere. The construction of landscape by US-American Laypersons. *Geographische Zeitschrift, 103*(3), 151–170.

Hokema, D. (2018). Was ist und was kann transdisziplinäre Landschaftsforschung? In K. Berr (Ed.), *Transdisziplinäre. Landschaftsforschung Grundlagen und Perspektiven* (pp. 25–40). Wiesbaden: Springer VS.

Homann, K., & Suchanek, A. (2005). *Ökonomik. Eine Einführung* (2., überarbeitete ed.). Tübingen: Mohr Siebeck.

Homer. (1994). *Odyssee*. Augsburg: Weltbild.

Honneth, A. (1992). *Kampf um Anerkennung. Zur moralischen Grammatik sozialer Konflikte*. Frankfurt a. M.: Suhrkamp.

Howard, P. J. (2011). *An Introduction to Landscape*. Farnham: Routledge.

Huber, A. (1999). *Heimat in der Postmoderne*. Zürich: Seismo.

Hubig, C. (2013). Dialektik des Entwerfens. Entwurfswissenschaft als Reflexion. In S. Ammon, & E. M. Froschauer (Eds.), *Wissenschaft Entwerfen. Vom forschenden Entwerfen zur Entwurfsforschung der Architektur* (eikones, pp. 267–285). München: Fink.

Hugill, P. J. (1995). *Upstate. Arcadia Landscape, aesthetics, and the triumph of social differentiation in America*. Lanham: Rowman & Littlefield.

Hülz, M., Kühne, O., & Weber, F. (Eds.). (2019). *Heimat. Ein vielfältiges Konstrukt*. Wiesbaden: Springer VS.

Hume, D. (1978). *Ein Traktat über die menschliche Natur. Buch II. Über die Affekte Buch III. Über Moral* (Unveränderter Nachdruck der 1. Auflage von 1906 (Buch 2 und 3)). Hamburg: Meiner.

Hunziker, M., & Kienast, F. (1999). Potential impacts of changing agricultural activities on scenic beauty – A prototypical technique for automated rapid assessment. *Landscape Ecology, 14*(2), 161–176. https://doi.org/10.1023/A:1008079715913.

Hupke, K.-D. (2015). *Naturschutz. Ein kritischer Ansatz*. Berlin: Springer Spektrum.

Hupke, K.-D. (2019a). Landschaftskonflikte um Naturschutzfragen: Der Naturschutz als schwächster der konkurrierenden Akteure? In K. Berr & C. Jenal (Eds.), *Landschaftskonflikte* (pp. 241–246). Wiesbaden: Springer VS.

Hupke, K.-D. (2019b). Naturschutz. In O. Kühne, F. Weber, K. Berr, & C. Jenal (Eds.), *Handbuch Landschaft* (pp. 479–487). Wiesbaden: Springer VS.

Hupke, K.-D. (2019c). Naturschutz und Heimat. In M. Hülz, O. Kühne, & F. Weber (Eds.), *Heimat Ein vielfältiges Konstrukt* (pp. 203–212). Wiesbaden: Springer VS.

Hüppauf, B. (2007). Heimat – die Wiederkehr eines verpönten Wortes Ein Populärmythos im Zeitalter der Globalisierung. In G. Gebhard, O. Geisler, & S. Schröter (Eds.), *Heimat. Konturen und Konjunkturen eines umstrittenen Konzepts* (pp. 109–140). Transcript: Bielefeld.

Husserl, E. (1954). *Die Krisis der europäischen Wissenschaften und die transzendentale Phänomenologie. Eine Einleitung in die phänomenologische Philosophie* (Husserliana, Vol. 6). Den Haag: Martinus Nijhoff (Herausgegeben von Walter Biemel).

Illing, F. (2006). *Kitsch, Kommerz und Kult Soziologie des schlechten Geschmacks*. Konstanz: UVK.

Ipsen, D., Reichhardt, U., Schuster, S., Wehrle, A., & Weichler, H. (2003). *Zukunft Landschaft. Bürgerszenarien zur Landschaftsentwicklung* (Arbeitsberichte des Fachbereichs Architektur, Stadtplanung, Landschaftsplanung, Vol. 153). Kassel: Selbstverlag.

Jackson, J. B. (1984). *Discovering the Vernacular Landscape*. New Haven: Yale University Press.

Jackson, J. B. (1997). Landscape Revisions. In H. L. Horowitz (Ed.), *Landscape in sight Looking at America* (pp. 333–371). London: Yale University Press.

Jaeschke, W. (2003). *Hegel-Handbuch. Leben – Werk – Schule*. Stuttgart: J. B. Metzler.

Jahn, T. (2008). Transdisziplinarität in der Forschungspraxis. In M. Bergmann & E. Schramm (Eds.), *Transdisziplinäre Forschung. Integrative Forschungsprozesse verstehen und bewerten* (pp. 21–37). New York: Campus.

Janich, P. (2015). *Handwerk und Mundwerk. Über das Herstellen von Wissen*. München: Beck.

Jauß, H. R. (1982). *Ästhetische Erfahrung und literarische Hermeneutik*. Frankfurt a. M.: Suhrkamp.

Jenal, C. (2019). *„Das ist kein Wald, Ihr Pappnasen!“ – Zur sozialen Konstruktion von Wald. Perspektiven von Landschaftstheorie und Landschaftspraxis*. Wiesbaden: Springer VS.

Jenal, C., & Schönwald, A. (2019). Wild drauflos? Wildniskonflikte im Fokus. In K. Berr & C. Jenal (Eds.), *Landschaftskonflikte* (pp. 375–388). Wiesbaden: Springer VS.

Jens, W. (1985). Nachdenken über Heimat Fremde und Zuhause im Spiegel deutscher Poesie. In H. Bienek (Ed.), *Heimat Neue Erkundungen eines alten Themas* (pp. 14–26). München: Hanser.

Jessel, B. (1995). Dimensionen des Landschaftsbegriffs. In Bayerische Akademie für Naturschutz und Landschaftspflege (Ed.), *Vision Landschaft 2020. Von der historischen Kulturlandschaft zur Landschaft von morgen* (pp. 7–10). Laufen: Druckerei Grauer.

Jessel, B. (2005). Landschaft. In E.-H. Ritter (Ed.), *Handwörterbuch der Raumordnung* (4th ed., pp. 579–586). Hannover: Selbstverlag.

Jones, M., & Daugstad, K. (1997). Usages of the "cultural landscape" concept in Norwegian and Nordic landscape administration. *Landscape Research, 22*(3), 267–281. https://doi.org/10.1080/01426399708706515.

Jungert, M., Romfeld, E., Sukopp, T., & Voigt, U. (Eds.). (2010). *Interdisziplinarität. Theorie, Praxis, Probleme.* Darmstadt: WBG.

Kant, I. (1993). *Kritik der Urteilskraft* (Die drei Kritiken: Jubiläumsausgabe anläßlich des 125-jährigen Bestehens der Philosophischen Bibliothek, Vol. 3). Hamburg: Meiner.

Kaufmann, S. (2005). *Soziologie der Landschaft.* Wiesbaden: VS Verlag.

Kazig, R. (2007). Atmosphären – Konzept für einen nicht repräsentationellen Zugang zum Raum. In C. Berndt & R. Pütz (Eds.), *Kulturelle Geographien. Zur Beschäftigung mit Raum und Ort nach dem Cultural Turn* (pp. 167–187). Bielefeld: transcript.

Kazig, R. (2013). Landschaft mit allen Sinnen – Zum Wert des Atmosphärenbegriffs für die Landschaftsforschung. In D. Bruns & O. Kühne (Eds.), *Landschaften: Theorie, Praxis und internationale Bezüge* (pp. 221–232). Schwerin: Oceano Verlag.

Kazig, R. (2019a). Atmosphären und Landschaft. In O. Kühne, F. Weber, K. Berr, & C. Jenal (Eds.), *Handbuch Landschaft* (pp. 453–460). Wiesbaden: Springer VS.

Kazig, R. (2019b). Für ein alltagsästhetisches Verständnis von Heimat. In M. Hülz, O. Kühne, & F. Weber (Eds.), *Heimat. Ein vielfältiges Konstrukt* (pp. 87–97). Wiesbaden: Springer VS.

Kianicka, S., Buchecker, M., Hunziker, M., & Müller-Böker, U. (2006). Locals' and tourists' sense of place. A case study of a Swiss Alpine Village. *Mountain Research and Development, 26*(1), 55–63. https://doi.org/10.1659/0276-4741(2006)026[0055:LATSOP]2.0.CO;2.

Kirchhoff, T. (2017). Landschaft. In T. Kirchhoff, N. C. Karafyllis, D. Evers, B. Falkenburg, M. Gerhard, G. Hartung, et al. (Eds.), *Naturphilosophie Ein Lehr- und Studienbuch* (pp. 152–158). Tübingen: Mohr Siebeck.

Kirchhoff, T. (2019). Ökosystemdienstleistungen. In O. Kühne, F. Weber, K. Berr, & C. Jenal (Eds.), *Handbuch Landschaft* (pp. 807–822). Wiesbaden: Springer VS.

Kirchhoff, T., & Trepl, L. (2009a). Landschaft, Wildnis, Ökosystem: zur kulturbedingten Vieldeutigkeit ästhetischer, moralischer und theoretischer Naturauffassungen. Einleitender Überblick. In T. Kirchhoff, & L. Trepl (Eds.), *Vieldeutige Natur. Landschaft, Wildnis und Ökosystem als kulturgeschichtliche Phänomene* (Sozialtheorie, pp. 13–68). Bielefeld: transcript.

Kirchhoff, T., & Trepl, L. (Eds.). (2009b). *Vieldeutige Natur. Landschaft, Wildnis und Ökosystem als kulturgeschichtliche Phänomene* (Sozialtheorie). Bielefeld: transcript.

Klages, L. (2013 [1913]). *Mensch und Erde – Ein Denkanstoß.* Berlin: Matthes & Seitz.

Klein, J. (2005). „Grundwortschatz" der Demokratie. In J. Kilian (Ed.), *Sprache und Politik. Deutsch im demokratischen Staat* (Thema deutsch, Vol. 6, pp. 128–140). Mannheim: Dudenverlag.

Klein, J. T. (2014). Discourses of transdisciplinarity: Looking back to the future. *Futures, 63,* 68–74. https://doi.org/10.1016/j.futures.2014.08.008.

Kluge, F. (1975). *Etymologisches Wörterbuch der deutschen Sprache* ((21, unveränderte Auflage ed.). Berlin: De Gruyter.

Kneer, G. (2009). Akteur-Netzwerk-Theorie. In G. Kneer & M. Schroer (Eds.), *Handbuch Soziologische Theorien* (pp. 19–39). Wiesbaden: VS Verlag.

Kniffen, F. (1965). Folk housing: Key to diffusion. *Annals of the Association of American Geographers, 55*(4), 549–576. https://doi.org/10.1111/j.1467-8306.1965.tb00535.x.

Kocka, J. (Ed.). (1987). *Interdisziplinarität. Praxis, Herausforderung, Ideologie.* Frankfurt a. M.: Suhrkamp.

Kook, K. (2008). Zum Landschaftsverständnis von Kindern: Aussichten – Ansichten – Einsichten. In R. Schindler, J. Stadelbauer, & W. Konold (Eds.), *Points of View Landschaft verstehen – Geographie und Ästhetik, Energie und Technik* (pp. 107–124). Freiburg: modo Verlag.

Körner, S. (2005). Landschaft und Raum im Heimat- und Naturschutz. In M. Weingarten (Ed.), *Strukturierung von Raum und Landschaft. Konzepte in Ökologie und der Theorie gesellschaftlicher Naturverhältnisse* (pp. 107–117). Westfälisches Dampfboot: Münster.

Körner, S. (2005b). *Natur in der urbanisierten Landschaft. Ökologie, Schutz und Gestaltung* (Zwischenstadt, vol. 4). Wuppertal: Müller + Busmann.

Körner, S. (2006a). Eine neue Landschaftstheorie? Eine Kritik am Begriff „Landschaft Drei". *Stadt + Grün, 10/2006*, 18–25.

Körner, S. (2006). Heimatschutz, Naturschutz und Landschaftsplanung. In Institut für Landschaftsarchitektur und Umweltplanung – Technische Universität Berlin (Ed.), *Perspektive Landschaft* (pp. 131–142). Berlin: wvb Wissenschaftlicher Verlag Berlin.

Körner, S. (2017). Aneignung in der Freiraumplanung – Eine ideengeschichtliche und planungstheoretische Einordnung. In T. E. Hauck, S. Hennecke, & S. Körner (Eds.), *Aneignung urbaner Freiräume. Ein Diskurs über städtischen Raum* (pp. 129–150). Bielefeld: transcript.

Körner, S., & Eisel, U. (2003). Naturschutz als kulturelle Aufgabe – Theoretische Rekonstruktrion und Anregungen für eine inhaltliche Erweiterung. In S. Körner, A. Nagel, & U. Eisel (Eds.), *Naturschutzbegründungen* (pp. 5–49). Bonn-Bad Godesberg: Selbstverlag.

Körner, S., & Eisel, U. (2006). Nachhaltige Landschaftsentwicklung. In D. D. Genske (Ed.), *Fläche – Zukunft – Raum. Strategien und Instrumente für Regionen im Umbruch* (Schriftenreihe der Deutschen Gesellschaft für Geowissenschaften, Vol. 37, pp. 45–60). Hannover: Deutsche Gesellschaft für Geowissenschaften.

Körner, S., Eisel, U., & Nagel, A. (2003). Heimat als Thema des Naturschutzes. Anregungen für eine sozio-kulturelle Erweiterung. *Natur und Landschaft, 78*(9/10), 382–389.

Kortländer, B. (1977). Die Landschaft in der Literatur des ausgehenden 18. und beginnenden 19. Jahrhunderts. In A. Hartlieb von Wallthor, & H. Quirin (Eds.), *„Landschaft" als interdisziplinäres Forschungsproblem. Vorträge und Diskussionen des Kolloquiums am 7./8. November 1975 in Münster*. Münster: Aschendorff.

Koselleck, R. (Ed.). (1979). *Historische Semantik und Begriffsgeschichte*. Stuttgart: Klett-Cotta.

Kost, S. (2017). Raumbilder und Raumwahrnehmung von Jugendlichen. In O. Kühne, H. Megerle, & F. Weber (Eds.), *Landschaftsästhetik und Landschaftswandel* (pp. 69–85). Wiesbaden: Springer VS.

Krebs, S., & Seifert, M. (2012). *Landschaft quer Denken. Theorien – Bilder – Formationen* (Schriften zur sächsischen Geschichte und Volkskunde, Vol. 39). Leipzig: Leipziger Universitäts-Verlag.

Kristeller, P. O. (1980). *Humanismus und Renaissance II. Philosophie, Bildung und Kunst*. München: Fink.

Krüger, R. (1972). *Das Zeitaler der Empfindsamkeit. Kunst und Kulter des späten 18. Jahrhunderts in Deutschland*. Leipzig: Koehler & Amelang.

Kühne, O. (2006). *Landschaft in der Postmoderne. Das Beispiel des Saarlandes*. Wiesbaden: DUV.

Kühne, O. (2008a). Die Sozialisation von Landschaft – sozialkonstruktivistische Überlegungen, empirische Befunde und Konsequenzen für den Umgang mit dem Thema Landschaft in Geographie und räumlicher Planung. *Geographische Zeitschrift, 96*(4), 189–206.

Kühne, O. (2008b). *Distinktion – Macht – Landschaft. Zur sozialen Definition von Landschaft*. Wiesbaden: VS Verlag.

Kühne, O. (2008c). Kritische Geographie der Machtbeziehungen – konzeptionelle Überlegungen auf der Grundlage der Soziologie Pierre Bourdieus. *geographische revue, 10*(2), 40–50.

Kühne, O. (2008d). Landschaft und Kitsch – Anmerkungen zu impliziten und expliziten Landschaftsvorstellungen. *Naturschutz und Landschaftsplanung, 44*(12), 403–408.

Kühne, O. (2009a). Heimat und Landschaft – Zusammenhänge und Zuschreibungen zwischen Macht und Mindermacht. Überlegungen auf sozialkonstruktivistischer Grundlage. *Stadt + Grün, 9*, 17–22.

Kühne, O. (2009b). Landschaft und Heimat – Überlegungen zu einem geographischen Amalgam. *Berichte zur deutschen Landeskunde, 83*(3), 223–240.

Kühne, O. (2011). Heimat und sozial nachhaltige Landschaftsentwicklung. *Raumforschung und Raumordnung, 69*(5), 291–301. https://doi.org/10.1007/s13147-011-0108-0.

Kühne, O. (2013). *Landschaftstheorie und Landschaftspraxis. Eine Einführung aus sozialkonstruktivistischer Perspektive.* Wiesbaden: Springer VS.

Kühne, O. (2014a). Das Konzept der Ökosystemdienstleistungen als Ausdruck ökologischer Kommunikation. Betrachtungen aus der Perspektive Luhmannscher Systemtheorie. *Naturschutz und Landschaftsplanung, 46*(1), 17–22.

Kühne, O. (2014b). Die intergenerationell differenzierte Konstruktion von Landschaft. *Naturschutz und Landschaftsplanung, 46*(10), 297–302.

Kühne, O. (2014c). „Im Wald, da sind die …“ Was eigentlich heute? Zur sozialen Bedeutung von Wald. Erste Ergebnisse einer Langzeitstudie. *Naturschutz im Saarland, 44*(2), 20–21.

Kühne, O. (2014d). Landschaft und Macht: von Eigenlogiken und Ästhetiken in der Raumentwicklung. *Ausdruck und Gebrauch, 12*, 151–172.

Kühne, O. (2015a). Historical developments: The evolution of the concept of landscape in German Linguistic Areas. In D. Bruns, O. Kühne, A. Schönwald, & S. Theile (Eds.), *Landscape culture – culturing landscapes. The differentiated construction of landscapes* (pp. 43–52). Wiesbaden: Springer VS.

Kühne, O. (2015b). The streets of Los Angeles: Power and the infrastructure landscape. *Landscape Research, 40*(2), 139–153. https://doi.org/10.1080/01426397.2013.788691.

Kühne, O. (2017). Der intergenerationelle Wandel landschaftsästhetischer Vorstellungen – eine Betrachtung aus sozialkonstruktivistischer Perspektive. In O. Kühne, H. Megerle, & F. Weber (Eds.), *Landschaftsästhetik und Landschaftswandel* (pp. 53–67). Wiesbaden: Springer VS.

Kühne, O. (2018a). ‚Neue Landschaftskonflikte‘ – Überlegungen zu den physischen Manifestationen der Energiewende auf der Grundlage der Konflikttheorie Ralf Dahrendorfs. In O. Kühne & F. Weber (Eds.), *Bausteine der Energiewende* (pp. 163–186). Wiesbaden: Springer VS.

Kühne, O. (2018b). Der doppelte Landschaftswandel. Physische Räume, soziale Deutungen, Bewertungen. *Nachrichten der ARL, 48*(*1*), 14–17. https://shop.arl-net.de/media/direct/pdf/nachrichten/2018-1/NR1-18K%C3%BChneS14-17online.pdf. Accessed 8 Oct 2018.

Kühne, O. (2018c). Die Moralisierung von Landschaft – Überlegungen zu einer problematischen Kommunikation aus Sicht der Luhmannschen Systemtheorie. In S. Hennecke, H. Kegler, K. Klaczynski, & D. Münderlein (Eds.), *Diedrich Bruns wird gelehrt haben. Eine Festschrift* (pp. 115–121). Kassel: Kassel University Press.

Kühne, O. (2018d). *Landscape and power in geographical space as a social-aesthetic construct.* Dordrecht: Springer International Publishing.

Kühne, O. (2018e). *Landschaft und Wandel. Zur Veränderlichkeit von Wahrnehmungen.* Wiesbaden: Springer VS.

Kühne, O. (2018f). *Landschaftstheorie und Landschaftspraxis. Eine Einführung aus sozialkonstruktivistischer Perspektive* (2., aktualisierte und überarbeitete ed.). Wiesbaden: Springer VS.

Kühne, O. (2018g). Macht, Herrschaft und Landschaft: Landschaftskonflikte zwischen Dysfunktionalität und Potenzial. Eine Betrachtung aus Perspektive der Konflikttheorie Ralf Dahrendorfs. In K. Berr (Ed.), *Transdisziplinäre Landschaftsforschung. Grundlagen und Perspektiven* (pp. 155–170). Wiesbaden: Springer VS.

Kühne, O. (2018h). Postmodernisierung und Großschutzgebiete – Überlegungen zu Natur, Raum und Planung aus sozialkonstruktivistischer Perspektive. In F. Weber, F. Weber, & C. Jenal (Eds.), Wohin des Weges? Regionalentwicklung in Großschutzgebieten (Arbeitsberichte der ARL, Vol. 21, pp. 44–55). Hannover: Selbstverlag.

Kühne, O. (2019a). Die Produktivität von Landschaftskonflikten – Möglichkeiten und Grenzen auf Grundlage der Konflikttheorie Ralf Dahrendorfs. In K. Berr & C. Jenal (Eds.), *Landschaftskonflikte* (pp. 37–49). Wiesbaden: Springer VS.

Kühne, O. (2019b). Die Sozialisation von Landschaft. In O. Kühne, F. Weber, K. Berr, & C. Jenal (Eds.), *Handbuch Landschaft* (pp. 301–312). Wiesbaden: Springer VS.

Kühne, O. (2019c). *Landscape theories. A brief introduction.* Wiesbaden: Springer VS.

Kühne, O. (2019d). Phänomenologische Landschaftsforschung. In O. Kühne, F. Weber, K. Berr, & C. Jenal (Eds.), *Handbuch Landschaft* (pp. 135–144). Wiesbaden: Springer VS.

Kühne, O. (2019e). Sich abzeichnende theoretische Perspektiven für die Landschaftsforschung: Neopragmatismus, Akteur-Netzwerk-Theorie und Assemblage-Theorie. In O. Kühne, F. Weber, K. Berr, & C. Jenal (Eds.), *Handbuch Landschaft* (pp. 153–162). Wiesbaden: Springer VS.

Kühne, O. (2019f). Sozialkonstruktivistische Landschaftstheorie. In O. Kühne, F. Weber, K. Berr, & C. Jenal (Eds.), *Handbuch Landschaft* (pp. 69–79). Wiesbaden: Springer VS.

Kühne, O. (2019g). Vom ‚Bösen‘ und ‚Guten‘ in der Landschaft – das Problem moralischer Kommunikation im Umgang mit Landschaft und ihren Konflikten. In K. Berr & C. Jenal (Eds.), *Landschaftskonflikte* (pp. 131–142). Wiesbaden: Springer VS.

Kühne, O., & Spellerberg, A. (2010). *Heimat und Heimatbewusstsein in Zeiten erhöhter Flexibilitätsanforderungen. Empirische Untersuchungen im Saarland.* Wiesbaden: VS Verlag für Sozialwissenschaften.

Kühne, O., & Weber, F. (2019a). Landschaft und Heimat – argumentative Verknüpfungen durch Bürgerinitiativen im Kontext des Stromnetz- und des Windkraftausbaus. In M. Hülz, O. Kühne, & F. Weber (Eds.), *Heimat. Ein vielfältiges Konstrukt* (pp. 163–178). Wiesbaden: Springer VS.

Kühne, O., & Weber, F. (2019b). Postmoderne Zugriffe und Differenzierungen von Stadt und Land(schaft): Stadtlandhybride, räumliche Pastiches und URFSURBS. In O. Kühne, F. Weber, K. Berr, & C. Jenal (Eds.), *Handbuch Landschaft* (pp. 755–770). Wiesbaden: Springer VS.

Kühne, O., Jenal, C., & Schönwald, A. (2013). Wie konstruiert der Mensch Landschaft? Zum Verhältnis von Landschaft, Heimat und Identität sowie Hinweise für deren Bedeutung für die räumliche Planung. In MULEWF (Ed.), *Landentwicklung und Ländliche Bodenordnung* (Landentwicklung und ländliche Bodenordnung – Nachrichtenblatt, Vol. 54, pp. 13–28). Mainz: Selbstverlag.

Kühne, O., Jenal, C., & Currin, A. (BMUB, BfN, & NABU Saarland, Eds.). (2014). Längsschnittstudie zur Wahrnehmung von Alt- und Totholz sowie zur symbolischen Konnotation von Wald. Zwischenbericht Phase 1. http://wertvoller-wald.de/uploads/media/ ZwischenberichtSozialwissenschaftlicheEvaluation.pdf. Accessed 3 May 2017.

Kühne, O., Weber, F., & Jenal, C. (2018). *Neue Landschaftsgeographie. Ein Überblick* (Essentials). Wiesbaden: Springer VS.

Kühne, O., Weber, F., & Jenal, C. (2019). Neue Landschaftsgeographie. In O. Kühne, F. Weber, K. Berr, & C. Jenal (Eds.), *Handbuch Landschaft* (pp. 119–134). Wiesbaden: Springer VS.

Kühne, O., Weber, F., & Berr, K. (2019). The productive potential and limits of landscape conflicts in light of Ralf Dahrendorf's conflict theory. *Società Mutamento Politica, 10(19)*, im Satz.

Küster, H. (2009). *Schöne Aussichten. Kleine Geschichte der Landschaft.* München: Beck.

Küster, H. (2012a). Arkadien als halboffene Weidelandschaft. *Merkur, 66(758)*, 651–656.

Küster, H. (2012b). *Die Entdeckung der Landschaft. Einführung in eine neue Wissenschaft.* München: C.H. Beck.

Laclau, E. (1996). *Emancipation(s)*. London: Verso.

Laclau, E., & Mouffe, C. (1985). *Hegemony and Socialist Strategy. Towards a Radical Democratic Politics*. London: Verso.

Laitko, H. (2018). *Grenzüberschreitungen*. Festvortrag, gehalten anlässlich des 10-jährigen Bestehens des Leibniz-Instituts für interdisziplinäre Studien, Berlin-Adlershof. https://leibniz-institut.de/archiv/laitko080712.pdf. Accessed 4 Dec 2018.

Langen, A. (1975 [1953]). Verbale Dynamik in der dichterischen Landschaftsschilderung des 18. Jahrhunderts (1948/49). In A. Ritter (Ed.), *Landschaft und Raum in der Erzählkunst* (Wege der Forschung, Vol. 418, pp. 112–191). Darmstadt: WBG.

Latour, B. (1998). *Wir sind nie modern gewesen. Versuch einer symmetrischen Anthropologie*. Frankfurt a. M.: Suhrkamp.

Latour, B. (2002 [1999]). *Die Hoffnung der Pandora. Untersuchungen zur Wirklichkeit der Wissenschaft*. Frankfurt a. M.: Suhrkamp.

Lautensach, H. (1973 [1938]). Über die Erfassung und Abgrenzung von Landschaftsräumen. In K. Paffen (Ed.), *Das Wesen der Landschaft* (Wege der Forschung, Vol. 39, pp. 20–38). Darmstadt: WBG.

Lehmann, H. (1968). *Formen landschaftlicher Raumerfahrung im Spiegel der bildenden Kunst*. Erlangen: Selbstverlag der FGG.

Lehmann, A. (1996). Wald als „Lebensstichwort". Zur biographischen Bedeutung der Landschaft, des Naturerlebnisses und des Naturbewußtseins. *BIOS, 9*(2), 143–154.

Lehmann, A. (2001a). Landschaftsbewusstsein. Zur gegenwärtigen Wahrnehmung natürlicher Ensembles. In R. W. Brednich, A. Schneider, & U. Werner (Eds.), *Natur – Kultur. Volkskundliche Perspektiven auf Mensch und Umwelt* (pp. 147–154). Münster: Waxmann.

Lehmann, A. (2001b). Mythos Deutscher Wald. *Der Bürger im Staat, 51*(1), 4–9.

Lehmann, A. (2004). Mythos Wald. *politische ökologie, 22*(89), 12–16.

Lehmann, A. (2010). Der deutsche Wald. Kulturmuster und Identitätssymbol. In O. Depenheuer, & B. Möhring (Eds.), *Waldeigentum. Dimensionen und Perspektiven* (Bibliothek des Eigentums, Vol. 8, pp. 3–19). Berlin: Springer.

Leibenath, M. (2014). Landschaftsbewertung im Spannungsfeld von Expertenwissen, Politik und Macht. *UVP-report, 28*(2), 44–49. https://www2.ioer.de/recherche/pdf/2014leibenathuvp-report.pdf. Accessed 26 Jan 2017.

Leibenath, M. (2015). Landschaften und Macht. In S. Kost & A. Schönwald (Eds.), *Landschaftswandel – Wandel von Machtstrukturen* (pp. 17–26). Wiesbaden: Springer VS.

Leibenath, M., & Gailing, L. (2012). Semantische Annäherung an „Landschaft" und „Kulturlandschaft". In W. Schenk, M. Kühn, M. Leibenath, & S. Tzschaschel (Eds.), Suburbane Räume als Kulturlandschaften (Forschungs- und Sitzungsberichte, Vol. 236, pp. 58–79). Hannover: Selbstverlag.

Leibenath, M., & Lintz, G. (2018). Streifzug mit Michel Foucault durch die Landschaften der Energiewende: Zwischen Government, Governance und Gouvernementalität. In O. Kühne & F. Weber (Eds.), *Bausteine der Energiewende* (pp. 91–107). Wiesbaden: Springer VS.

Leibenath, M., & Otto, A. (2012). Diskursive Konstituierung von Kulturlandschaft am Beispiel politischer Windenergiediskurse in Deutschland. *Raumforschung und Raumordnung, 70*(2), 119–131. https://doi.org/10.1007/s13147-012-0148-0.

Leser, H. (2019). Landschaftsökologie. In O. Kühne, F. Weber, K. Berr, & C. Jenal (Eds.), *Handbuch Landschaft* (pp. 181–191). Wiesbaden: Springer VS.

Leser, H. (2019). Stadtökologie. In O. Kühne, F. Weber, K. Berr, & C. Jenal (Eds.), *Handbuch Landschaft* (pp. 193–205). Wiesbaden: Springer VS.

Löfgren, O. (2002). *On holiday. A history of vacationing*. Berkeley: University of California Press.

Loibl, M. C. (2005). *Spannungen in Forschungsteams. Hintergründe und Methoden zum konstruktiven Abbau von Konflikten in inter- und transdisziplinären Projekten.* Heidelberg: Carl-Auer-Systeme.

Lorenzen, P. (1974). Interdisziplinäre Forschung und infradisziplinäres Wissen. In P. Lorenzen (Ed.), *Konstruktive Wissenschaftstheorie.* Suhrkamp: Frankfurt a. M.

Lorris, G. de, & Meung, J. de. (1976–79). *Der Rosenroman.* München: Fink.

Luhmann, N. (1984). *Soziale Systeme. Grundriß einer allgemeinen Theorie.* Frankfurt a. M.: Suhrkamp.

Luhmann, N. (1986). *Ökologische Kommunikation. Kann die moderne Gesellschaft sich auf ökologische Gefährdungen einstellen?.* Opladen: Westdeutscher Verlag.

Luhmann, N. (2017). *Systemtheorie der Gesellschaft.* Berlin: Suhrkamp.

Lützeler, H. (1950). Vom Wesen der Landschaftsmalerei. In K. H. Bauer, L. Curtius, H. v. Einem, F. Ernst, H. Friedrich, W. Fucks, et al. (Eds.), *Studium Generale. Eitschrift für die Einheit der Wissenschaften im Zusammenhang ihrer Begriffsbildungen und Forschungsmethoden.* Dritter Jahrgang (pp. 210–232). Berlin, Heidelberg: Springer.

Magnus, A. (1867 [first around second half of the 13th century]). *Alberti Magni ex Ordine Praedicatorum De vegetabilibus libri VII, historiae naturalis pars XVIII.* Berlin: Reimer.

Makhzoumi, J. M. (2002). Landscape in the Middle East: An inquiry. *Landscape Research, 27*(3), 213–228. https://doi.org/10.1080/01426390220149494.

Makhzoumi, J. M. (2015). Borrowed or rooted? The discourse of 'Landscape' in the Arab Middle East. In D. Bruns, O. Kühne, A. Schönwald, & S. Theile (Eds.), *Landscape culture – Culturing landscapes. The differentiated construction of landscapes* (pp. 111–126). Wiesbaden: Springer VS.

Marg, S. (2019). „Deutschland – meine Heimat, meine Liebe.". Die Verhandlung von Heimat im Umfeld von, Pegida'. In M. Hülz, O. Kühne, & F. Weber (Eds.), *Heimat. Ein vielfältiges Konstrukt* (pp. 355–369). Wiesbaden: Springer VS.

Mathewson, K. (2009). Carl Sauer and his crititcs. In W. M. Denevan & K. Mathewson (Eds.), *Carl Sauer on culture and landscape. Readings and commentaries* (pp. 9–28). Baton Rouge: Louisiana State University Press.

Mattissek, A., & Wiertz, T. (2014). Materialität und Macht im Spiegel der Assemblage-Theorie: Erkundungen am Beispiel der Waldpolitik in Thailand. *Geographica Helvetica, 69*(3), 157–169.

Maturana, H. R. (1985 [1982]). *Erkennen: Die Organisation und Verkörperung von Wirklichkeit. Ausgewählte Arbeiten zur biologischen Epistemologie* (Wissenschaftstheorie, Wissenschaft und Philosophie, Vol. 19, 2. Aufl.). Braunschweig: Vieweg.

Mayntz, R. (2005). Governance Theory als fortentwickelte Steuerungstheorie? In G. F. Schuppert (Ed.), *Governance-Forschung. Vergewisserung über Stand und Entwicklungslinien* (pp. 11–20). Baden-Baden: Nomos.

Micheel, M. (2012). Alltagsweltliche Konstruktionen von Kulturlandschaft. *Raumforschung und Raumordnung, 70*(2), 107–117. https://doi.org/10.1007/s13147-011-0143-x.

Mitchell, D. (2008). New axioms for reading the landscape: Paying attention to political economy and social justice. In J. L. Wescoat, & D. M. Johnston (Eds.), *Political economies of landscape change. Places of integrative power* (The GeoJournal Library, Vol. 89, pp. 29–50). Dordrecht: Springer.

Mitchell, W. J. T. (2002). Introduction. In W. J. T. Mitchell (Ed.), *Landscape and power* (2nd ed., pp. 1–4). Chicago: University of Chicago Press.

Mitscherlich, A. (1965). *Die Unwirtlichkeit unserer Städte. Anstiftung zum Unfrieden.* Frankfurt a. M.: Suhrkamp.

Mittelstraß, J. (1998). Interdisziplinarität oder Transdisziplinarität? *Die Häuser des Wissens. Wissenschaftstheoretische Studien* (pp. 29–48). Suhrkamp: Frankfurt (Main).

Mittelstraß, J. (2004). „Begriffsgeschichte". In J. Mittelstraß (Ed.), *Enzyklopädie Philosophie und Wissenschaftstheorie* (1, A – G, pp. 270–271). Stuttgart: J. B. Metzler.

Mittelstraß, J. (2005). Methodische Transdisziplinarität. *Technikfolgenabschätzung – Theorie und Praxis, 14*(2), 18–23.

Motz, H. (1865). *Über die Empfindung der Naturschönheit bei den Alten.* Leipzig: Hirzel.

Mouffe, C. (2000). Deliberative Democracy or Agonistic Pluralism. Reihe Politikwissenschaft Political Science Series: 72. https://www.ihs.ac.at/publications/pol/pw72.pdf. Accessed 13 Oct 2017.

Mouffe, C. (2005). *On the political.* London: Routledge.

Mouffe, C. (2007). *Über das Politische. Wider die kosmopolitische Illusion.* Frankfurt a. M.: Suhrkamp.

Müller, G. (1977). Zur Geschichte des Wortes Landschaft. In A. Hartlieb von Wallthor & H. Quirin (Eds.), *„Landschaft" als interdisziplinäres Forschungsproblem. Vorträge und Diskussionen des Kolloquiums am 7./8. November 1975 in Münster* (pp. 3–13). Münster: Aschendorff.

Müller, E., & Schmieder, F. (2016). *Begriffsgeschichte und historische Semantik. Ein kritisches Kompendium.* Berlin: Suhrkamp.

Naess, A. (1973). The shallow and the deep, long-range ecology movement. A summary. *Inquiry: An Interdisciplinary Journal of Philosophy, 16*(1–4), 95–100. https://doi.org/10.1080/00201747308601682.

Nissen, U. (1998). *Kindheit, Geschlecht und Raum. Sozialisationstheoretische Zusammenhänge geschlechtsspezifischer Raumaneignung.* Weinheim: Beltz Juventa.

Nogué i Font, J. (1993). Toward a phenomenology of landscape and landscape experience: An example from Catalonia. In D. Seamon (Ed.), *Dwelling, seeing, and designing. Toward a phenomenological ecology* (pp. 159–180). Albany: State University of New York Press.

Nohl, W. (2016). Windkraftwerke sind keine Windmühlen. Warum moderne»Energie-Landschaften«nicht schön sind. In G. Etscheit (Ed.), *Geopferte Landschaften. Wie die Energiewende unsere Umwelt zerstört* (pp. 114–136). München: Heyne.

Olwig, K. (2008). The Jutland Ciper: Unlocking the meaning and power of a contested landscape. In M. Jones & K. Olwig (Eds.), *Nordic landscapes. Region and belonging on the northern edge of Europe* (pp. 12–52). Minneapolis: University of Minnesota Press; Published in cooperation with the Center for American Places.

Olwig, K. R. (1995). Reinventing common nature: Yosemite and Mount Rushmore – A meandering tale of a double nature. In W. Cronon (Ed.), *Uncommon ground. Toward reinventing nature* (pp. 379–408). New York: W. W. Norton & Company.

Olwig, K. R. (1996). Recovering the substantive nature of landscape. *Annals of the Association of American Geographers, 86*(4), 630–653.

Olwig, K. R. (2002). *Landscape, nature, and the body politic. From Britain's renaissance to America's new world.* Madison: University of Wisconsin Press.

Olwig, K. R. (2007). The practice of landscape‚Conventions' and the Just landscape: The case of the European landscape convention. *Landscape Research, 32*(5), 579–594. https://doi.org/10.1080/01426390701552738.

Oppel, A. (1884). *Landschaftskunde. Versuch einer Physiogonomik der gesamten Erdoberfläche in Skizzen, Charakteristiken und Schilderungen.* Breslau: Hirt.

Ott, K. (2007). „Heimat"-Argumente als Naturschutzbegründungen in Vergangenheit und Gegenwart. In Bundesamt für Naturschutz (Ed.), *Heimat und Naturschutz. Die Vilmer Thesen und ihre Kritiker* (Naturschutz und Biologische Vielfalt, Vol. 47, pp. 43–65). Bonn-Bad Godesberg: Selbstverlag.

Otto, A., & Leibenath, M. (2013). Windenergielandschaften als Konfliktfeld. Landschaftskonzepte, Argumentationsmuster und Diskurskoalitionen. In L. Gailing & M. Leibenath (Eds.), *Neue*

Energielandschaften – Neue Perspektiven der Landschaftsforschung (pp. 65–75). Wiesbaden: Springer VS.

Paffen, K. (Ed.). (1973a). *Das Wesen der Landschaft* (Wege der Forschung, Vol. 39). Darmstadt: WBG.

Paffen, K. (1973b). Der Landschaftsbegriff als Problemstellung (1953). In K. Paffen (Ed.), Das Wesen der Landschaft (Wege der Forschung, Vol. 39, pp. 71–112). Darmstadt: WBG.

Panofsky, E. (1975 [1936]). *Sinn und Deutung in der bildenden Kunst*. Köln: DuMont Schauberg.

Panofsky, E. (1980[1927]). Die Perspektive als ‚symbolische Form'. In H. Oberer, & E. Verheyen (Eds.), *Aufsätze zu Grundfragen der Kunstwissenschaft* (pp. 99–167). Berlin: Spiess.

Parsons, T. (1991 [1951]). *The Social System*. London: Routledge.

Perpeet, W. (1987). *Das Kunstschöne. Sein Ursprung in der italienischen Renaissance*. Freiburg (Breisgau): Alber.

Petrarca, F. (Ed.). (1995). *Die Besteigung des Mont Ventoux*. Stuttgart: Reclam.

Piechocki, R. (2006). Landschaft und Heimat. Zur Verdängung der kulturellen Dimensionen aus dem Naturschutz. In K. P. Wiemer (Ed.), *Dem Erbe verpflichtet. 100 Jahre Kulturlandschaftspflege im Rheinland. Festschrift zum 100-jährigen Bestehen des Rheinischen Vereins für Denkmalpflege und Landschaftsschutz* (pp. 321–337). Köln: Verlag des Rheinischen Vereins für Denkmalpflege und Landschaftsschutz.

Piechocki, R. (2007). Heimat – Begriffsentstehung und Begriffswandel. In Bundesamt für Naturschutz (Ed.), *Heimat und Naturschutz. Die Vilmer Thesen und ihre Kritiker* (Naturschutz und Biologische Vielfalt, Vol. 47, pp. 19–42). Bonn-Bad Godesberg: Selbstverlag.

Piechocki, R. (2010). *Landschaft – Heimat – Wildnis. Schutz der Natur – aber welcher und warum?*. München: Beck.

Piechocki, R., Eisel, U., Körner, S., Nagel, A., Wiersbinski, & Norbert. (2007). Die Vilmer Thesen zu „Heimat" und Naturschutz. In Bundesamt für Naturschutz (Ed.), *Heimat und Naturschutz. Die Vilmer Thesen und ihre Kritiker* (Naturschutz und Biologische Vielfalt, Vol. 47, pp. 9–18). Bonn-Bad Godesberg: Selbstverlag.

Piepmeier, R. (1980). Das Ende der ästhetischen Kategorie „Landschaft". Zu einem Aspekt neuzeitlichen Naturverhältnisses. *Westfälische Forschungen, 30*, 8–46.

Piltz, E. (2007). Verortung der Erinnerung Heimat und Raumerfahrung in Selbstzeugnissen der Frühen Neuzeit. In G. Gebhard, O. Geisler, & S. Schröter (Eds.), *Heimat Konturen und Konjunkturen eines umstrittenen Konzepts* (pp. 57–80). Bielefeld: transcript.

Plessner, H. (1981). *Die Stufen des Organischen und der Mensch. Einleitung in die philosophische Anthropologie* (Gesammelte Schriften, Vol. 4). Frankfurt a. M.: Suhrkamp.

Plinius Caecilius Secundus, G. (1990 [first around turn of the 1st to the 2nd century]. *Briefe. Lateinisch-deutsch = Epistularum libri decem* (Sammlung Tusculum, 6th ed.). München: Artemis-Verlag.

Plumpe, G. (1993). *Ästhetische Kommunikation der Moderne. Band 1: Von Kant bis Hegel*. Opladen: Westdeutscher Verlag.

Poerting, J., & Marquardt, N. (2019). Kritisch-geographische Perspektiven auf Landschaft. In O. Kühne, F. Weber, K. Berr, & C. Jenal (Eds.), *Handbuch Landschaft* (pp. 145–152). Wiesbaden: Springer VS.

Pöggeler, O. (1999). *Hegels Kritik der Romantik* (Philosophie an der Jahrtausendwende, Vol. 2). München: Wilhelm Fink.

Pöggeler, O. (2005). Hegel und Caspar David Friedrich. In A. Gethmann-Siefert, B. Collenberg-Plotnikov, & L. de Vos (Eds.), *Die geschichtliche Bedeutung der Kunst und die Bestimmung der Künste* (pp. 227–243). München: Fink.

Pohl, C., & Hirsch Hadorn, G. (2008). Gestaltung transdisziplinärer Forschung. *Sozialwissenschaften und Berufspraxis, 31*(2), 5–22.

Popper, K. R. (1963). *Conjectures and refutations. The growth of scientific knowledge*. London: Routledge & Kegan Paul.

Poser, H. (2004). Entwerfen als Lebensform – Elemente technischer Mobilität. In K. Kornwachs (Ed.), Technik – System – Verantwortung (Technikphilosophie, Vol. 10, pp. 561–575). Münster: LIT Verlag.

Potthast, T. (2010). Epistemisch-moralische Hybride und das Problem interdisziplinärer Urteilsbildung. In M. Jungert, E. Romfeld, T. Sukopp, & U. Voigt (Eds.), *Interdisziplinarität. Theorie, Praxis, Probleme* (pp. 173–191). Darmstadt: WBG.

Price, M., & Lewis, M. (1993). The reinvention of cultural geography. *Annals of the Association of American Geographers, 83*(1), 1–17. https://doi.org/10.1111/j.1467-8306.1993.tb01920.x.

Prominski, M. (2004). *Landschaft entwerfen. Zur Theorie aktueller Landschaftsarchitektur*. Berlin: Reimer.

Quante, M. (2008). *Einführung in die allgemeine Ethik* (Einführungen Philosophie, 3. Aufl.). Darmstadt: Wissenschaftliche Buchgesellschaft.

Quasten, H. (1997). Grundsätze und Methoden der Erfassung und Bewertung kulturhistorischer Phänomene der Kulturlandschaft. In W. Schenk, K. Fehn, & D. Denecke (Eds.), *Kulturlandschaftspflege. Beiträge der Geographie zur räumlichen Planung* (pp. 19–34). Berlin: Borntraeger.

Radkau, J. (2002). *Mensch und Natur in der Geschichte*. Leipzig: Klett Schulbuchverlag.

Reußwig, F. (2019). Heimat und politische Parteien. In M. Hülz, O. Kühne, & F. Weber (Eds.), *Heimat. Ein vielfältiges Konstrukt* (pp. 371–389). Wiesbaden: Springer VS.

Riedel, W. (1989). *„Der Spaziergang". Ästhetik der Landschaft und Geschichtsphilosophie der Natur bei Schiller*. Würzburg: Königshausen & Neumann.

Riehl, W. H. (1854). *Die Naturgeschichte des Volkes als Grundlage einer deutschen Social-Politik. Land und Leute* (Vol. 1). Stuttgart: Klett-Cotta.

Riehl, W. H. (1996). Das landschaftliche Auge. In G. Gröning & U. Herlyn (Eds.), *Landschaftswahrnehmung und Landschaftserfahrung* (Arbeiten zur sozialwissenschaftlich orientierten Freiraumplanung, pp. 144–162). Münster: LIT.

Ritter, J. (1974 [1963]). *Subjektivität. Sechs Aufsätze*. Frankfurt a. M.: Suhrkamp.

Ritter, A. (Ed.). (1975 [1953]). *Landschaft und Raum in der Erzählkunst* (Wege der Forschung, Vol. 418). Darmstadt: WBG.

Ritter, J. (1996). Landschaft. Zur Funktion des Ästhetischen in der modernen Gesellschaft. In G. Gröning & U. Herlyn (Eds.), *Landschaftswahrnehmung und Landschaftserfahrung* (Arbeiten zur sozialwissenschaftlich orientierten Freiraumplanung, pp. 28–68). Münster: LIT.

Rombach, H. (1980). *Phänomenologie des gegenwärtigen Bewußtseins*. Freiburg: Alber.

Roßmeier, A., Weber, F., & Kühne, O. (2018). Wandel und gesellschaftliche Resonanz – Diskurse um Landschaft und Partizipation beim Windkraftausbau. In O. Kühne & F. Weber (Eds.), *Bausteine der Energiewende* (pp. 653–679). Wiesbaden: Springer VS.

Roters, E. (1995). *Jenseits von Arkadien. Die romantische Landschaft*. Köln: DuMont.

Rudorff, E. (1994 [1897]). *Heimatschutz*. St. Goar: Reichl Verlag.

Safranski, R. (2007). *Romantik. Eine deutsche Affäre*. München: Hanser.

Sauder, G. (1974). *Empfindsamkeit Band I: Voraussetzungen und Elemente*. Stuttgart: Metzler.

Sauer, C. O. (1969 [1925]). The Morphology of Landscape. In J. Leighly (Ed.), *Land and life. A selection from the writings of Carl Ortwin Sauer* (pp. 19–53). Berkeley: University of California Press.

Schäfer, M. (2013). Inter- und Transdisziplinäre Nachhaltigkeitsforschung – Innovation durch Integration? In J. Rückert-John (Ed.), *Soziale Innovation und Nachhaltigkeit* (pp. 171–194). Wiesbaden: Springer Fachmedien.

Schemel, H.-J. (2004). Emotionaler Naturschutz – Zur Bedeutung von Gefühlen in naturschutzfachlichen Entscheidungsprozessen. *Natur und Landschaft, 79*(8), 371–378.

Schemel, H.-J., Laßberg, D. v., Meyer, G., Meyer, W., & Vielhaber, A. (Eds.). (2001). *Kommunikation und Umwelt im Tourismus. Empirische Grundlagen und Konzeptbausteine für ein nachhaltiges Verbraucherverhalten im Urlaub* (Umweltforschungsplan des Bundesministeriums für Umwelt, Naturschutz und Reaktorsicherheit, 2001, 2). Berlin: Schmidt.

Schenk, W. (2001a). Kulturlandschaft in Zeiten verschärfter Nutzungskonkurrenz: Genese, Akteure, Szenarien. In Akademie für Raumforschung und Landesplanung (Ed.), *Die Zukunft der Kulturlandschaft zwischen Verlust, Bewahrung und Gestaltung* (Forschungs- und Sitzungsberichte, Vol. 215, pp. 30–44). Hannover: Selbstverlag.

Schenk, W. (2001). Landschaft. In H. Beck, D. Geuenich, & H. Steuer (Eds.), *Reallexikon der Germanischen Altertumskunde* (Vol. 17, pp. 617–630). Berlin: De Gruyter.

Schenk, W. (2002). „Landschaft" und „Kulturlandschaft" – „getönte" Leitbegriffe für aktuelle Konzepte geographischer Forschung und räumlicher Planung. *Petermanns Geographische Mitteilungen, 146*(6), 6–13.

Schenk, W. (2006). Der Terminus „gewachsene Kulturlandschaft" im Kontext öffentlicher und raumwissenschaftlicher Diskurse zu „Landschaft" und Kulturlandschaft". In U. Matthiesen, R. Danielzyk, S. Heiland, & S. Tzschaschel (Eds.), *Kulturlandschaften als Herausforderung für die Raumplanung. Verständnisse – Erfahrungen – Perspektiven* (Forschungs- und Sitzungsberichte, Vol. 228, pp. 9–21). Hannover: Selbstverlag.

Schenk, W. (2011). *Historische Geographie* (Geowissen kompakt). Darmstadt: WBG.

Schenk, W. (2013). Landschaft als zweifache sekundäre Bildung – Historische Aspekte im aktuellen Gebrauch von Landschaft im deutschsprachigen Raum, namentlich in der Geographie. In D. Bruns & O. Kühne (Eds.), *Landschaften: Theorie, Praxis und internationale Bezüge* (pp. 23–36). Schwerin: Oceano Verlag.

Schenk, W. (2017). Landschaft. In L. Kühnhardt & T. Mayer (Eds.), *Bonner Enzyklopädie der Globalität. Band 1 und Band 2* (pp. 671–684). Wiesbaden: Springer VS.

Schenk, W., Fehn, K., & Denecke, D. (Eds.). (1997). *Kulturlandschaftspflege. Beiträge der Geographie zur räumlichen Planung.* Berlin: Borntraeger.

Schiller, F. (2004a). Der Spaziergang. In A. Meier (Ed.), *Sämtliche Werke. Band 1: Gedichte. Dramen 1* (pp. 228–234). München: dtv.

Schiller, F. (2004b). Über die ästhetische Erziehung des Menschen in einer Reihe von Briefen. In W. Riedel (Ed.), *Sämtliche Werke. Band 5: Erzählungen. Theoretische Schriften* (pp. 570–669). München: dtv.

Schiller, F. (2004c). Über naive und sentimentalische Dichtung. In W. Riedel (Ed.), *Sämtliche Werke. Band 5: Erzählungen. Theoretische Schriften* (pp. 694–780). München: dtv.

Schiller, F. (2004d). Was heißt und zu welchem Ende studiert man Universalgeschichte. In W. Riedel (Ed.), *Sämtliche Werke. Band 4: Historische Schriften* (pp. 749–767). München: dtv.

Schmidt, C., Hage, G., Hoppenstedt, A., Bruns, D., Kühne, O., Schuster, L., et al. (2018). *Landschaftsbild & Energiewende. Band 1: Grundlagen.* Bonn-Bad Godesberg: Bundesamt für Naturschutz.

Schneider, G. (1989). *Die Liebe zur Macht. Über die Reproduktion der Enteignung in der Landschaftspflege.* Kassel: Arbeitsgemeinschaft Freiraum und Vegetation.

Schneider, N. (2009). *Geschichte der Landschaftsmalerei. Vom Spätmittelalter bis zur Romantik.* Darmstadt: WBG.

Schneider, J. H. J., Sieferle, R. P., & Wils, J.-P. (Eds.). (1992). *Natur als Erinnerung? Annäherung an eine müde Diva* (Attempto-Disput). Tübingen: Attempto-Verlag.

Schultheiß, G. (2007). Alles Landschaft? Zur Konjunktur eines Begriffes in der Urbanistik. In U. Eisel & S. Körner (Eds.), *Landschaft in einer Kultur der Nachhaltigkeit. Band 2. Landschaftsgestaltung im Spannungsfeld zwischen Ästhetik und Nutzen* (Arbeitsberichte des Fachbereichs Architektur, Stadtplanung, Landschaftsplanung, Vol. 166, pp. 86–104). Kassel: Kassel University Press.

Schultz, H.-D. (1980). *Die deutschsprachige Geographie 1800–1970. Ein Beitrag zur Geschichte ihrer Methodologie.* Berlin: Geographisches Institut der Freien Universität.

Schultze-Naumburg, P. (1901–1917). *Die Kulturarbeiten. 9 Bände und 1 Ergänzungsband.* München: Callwey.

Schütz, A. (1960 [1932]). *Der sinnhafte Aufbau der sozialen Welt. Eine Einleitung in die Verstehende Soziologie* (2. Aufl.). Wien: Julius Springer (Original work published 1932).

Schütz, A. (1971 [1962]). *Gesammelte Aufsätze 1. Das Problem der Wirklichkeit.* Den Haag: Martinus Nijhoff.

Schütz, A. (1971). *Gesammelte Aufsätze 3. Studien zur phänomenologischen Philosophie.* Den Haag: Martinus Nijhoff.

Schütz, A., & Luckmann, T. (2003 [1975]). *Strukturen der Lebenswelt.* Konstanz: UTB.

Schwarzer, M. (2014). *Von Mondlandschaften zur Vision eines neuen Seenlandes. Der Diskurs über die Gestaltung von Tagebaubrachen in Ostdeutschland.* Wiesbaden: Springer VS.

Schweda, M. (2013). *Entzweiung und Kompensation. Joachim Ritters philosophische Theorie der modernen Welt.* Freiburg (Breisgau): Karl Alber.

Seel, M. (1996). *Eine Ästhetik der Natur* (Vol. 1231). Frankfurt a. M.: Suhrkamp.

Seidel, S. (Ed.). (1962). *Der Briefwechsel zwischen Friedrich Schiller und Wilhelm von Humboldt* (Vol. 2). Berlin: Aufbau-Verlag.

Seifert, M. (2012). Heimat in Bewegung. Zur Suche nach soziokultureller Identität in der Spätmoderne. In J. Klose, R. Lindner, & M. Seifert (Eds.), *Heimat heute. Reflexionen und Perspektiven* (pp. 15–34). Dresden: Thelem.

Sen, A. K. (1966). Hume's Law and Hare's Rule. *Philosophy, 41*(155), 75–79.

Shepard, P. (1967). *Man in the landscape. A historic view of the Esthetics of nature.* New York: Knopf.

Sieferle, R. P. (1985). Heimatschutz und das Ende der romantischen Utopie. *Arch +, 81*, (38–42).

Sieferle, R. P. (1986). Entstehung und Zerstörung der Landschaft. In M. Smuda (Ed.), *Landschaft* (pp. 238–265). Frankfurt a. M.: Suhrkamp.

Sieferle, R. P. (1997). *Rückblick auf die Natur. Eine Geschichte des Menschen und seiner Umwelt.* München: Luchterhand.

Sieverts, T. (1997). *Zwischenstadt. Zwischen Ort und Welt, Raum und Zeit, Stadt und Land* (Bauwelt Fundamente, Vol. 118). Braunschweig: Vieweg + Sohn.

Sieverts, T. (1998 [1997]). *Zwischenstadt. Zwischen Ort und Welt, Raum und Zeit, Stadt und Land* (Bauwelt Fundamente, Vol. 118, 2., durchgesehene und um ein Nachwort ergänzte Aufl.). Braunschweig: Vieweg + Sohn.

Sieverts, T. (2004). Sieben einfache Zugänge zum Begreifen und zum Umgang mit der Zwischenstadt. In N. Gestring, H. Glasauer, C. Hannemann, W. Petrowsky, & J. Pohlan (Eds.), *Jahrbuch StadtRegion 2003. Schwerpunkt: Urbane Regionen* (pp. 43–60). Opladen: Leske + Budrich.

Simmel, G. (1957 [1913]). *Brücke und Tür. Essays des Philosophen zur Geschichte, Religion, Kunst und Gesellschaft.* Stuttgart: K. F. Kohler Verlag.

Simmel, G. (1990). Philosophie der Landschaft. In G. Gröning & U. Herlyn (Eds.), *Landschaftswahrnehmung und Landschaftserfahrung. Texte zur Konstitution und Rezeption von Natur als Landschaft* (pp. 67–79). München: Minerva.

Simmel, G. (1995 [1902]). Der Bildrahmen. Ein ästhetischer Versuch. In R. Kramme, & O. Rammstedt (Eds.), *Aufsätze und Abhandlungen 1901–1908. Band 1.* Gesamtausgabe Band 7 (pp. 101–108). Frankfurt a. M.: Suhrkamp.

Sloterdijk, P. (2007). *Der ästhetische Imperativ. Schriften zur Kunst* (Fundus-Bücher, Vol. 166, 2., unveränderte Neuauflage). Hamburg: Philo & Philo Fine Arts.

Smuda, M. (Ed.). (1986). *Landschaft.* Frankfurt a. M.: Suhrkamp.

Snell, B. (1975[1945]). *Die Entdeckung des Geistes. Studien zur Entstehung des europäischen Denkens bei den Griechen.* Göttingen: Vandenhoeck & Ruprecht.

Spanier, H. (2006). Pathos der Nachhaltigkeit. Von der Schwierigkeit, „Nachhaltigkeit" zu kommunizieren. *Stadt + Grün, 12,* 26–33.

Spanier, H. (2008). Mensch und Natur – Reflexionen über unseren Platz in der Natur. In Bundesamt für Naturschutz (Ed.), *Naturschutz im Kontext einer nachhaltigen Entwicklung. Ansätze, Konzepte, Strategien* (Naturschutz und Biologische Vielfalt, Vol. 67, pp. 269–292). Bonn: Bundesamt für Naturschutz.

Spence, J. (1966). *Observations, Anecdotes, and Characters of Books and Men. Collected from conversation.* Oxford: Clarendon.

Spirn, A. W. (1998). *The Language of Landscape.* New Haven: Yale University Press.

Steingräber, E. (1985). *Zweitausend Jahre europäische Landschaftsmalerei.* München: Hirmer.

Steinmann, K. (1995). Nachwort. Grenzscheide zweier Welten – Petrarcas Besteigung des Mont Ventoux. In F. Petrarca (Ed.), *Die Besteigung des Mont Ventoux* (pp. 39–49). Stuttgart: Reclam.

Stemmer, B. (2016). *Kooperative Landschaftsbewertung in der räumlichen Planung. Sozialkonstruktivistische Analyse der Landschaftswahrnehmung der Öffentlichkeit.* Wiesbaden: Springer VS.

Stiens, G. (2009). *Gegen den Verfall lebensweltlicher Landschaften* (Beiträge zur Sozialästhetik, Vol. 9). Bochum: Projekt-Verlag.

Stierle, K. (1979). *Petrarcas Landschaften. Zur Geschichte ästhetischer Landschaftserfahrung.* Krefeld: Scherpe.

Stotten, R. (2013). Kulturlandschaft gemeinsam verstehen – Praktische Beispiele der Landschaftssozialisation aus dem Schweizer Alpenraum. *Geographica Helvetica, 68*(2), 117–127. https://doi.org/10.5194/gh-68-117-2013.

Stotten, R. (2019). Kulturlandschaft als Ausdruck von Heimat der bäuerlichen Gesellschaft. In M. Hülz, O. Kühne, & F. Weber (Eds.), *Heimat. Ein vielfältiges Konstrukt* (pp. 149–162). Wiesbaden: Springer VS.

Straßburg, G. v. (1991). *Tristan und Isolde.* Frankfurt a. M.: Insel.

Stuhlmann-Laeisz, R. (1983). *Das Sein-Sollen-Problem. Eine modallogische Studie* (Problemata, Vol. 96). Stuttgart: Frommann-Holzboog.

Sukopp, T. (2010). Interdisziplinarität und Transdisziplinarität. In M. Jungert, E. Romfeld, T. Sukopp, & U. Voigt (Eds.), *Interdisziplinarität. Theorie, Praxis, Probleme* (pp. 13–29). Darmstadt: WBG.

Taylor, K., & Xu, Q. (2018). Challenging landscape Eurocentrism: An Asian perspective. In P. Howard, I. H. Thompson, E. Waterton, & M. Atha (Eds.), *The Routledge companion to landscape studies* (2nd ed., pp. 311–328). Abingdon: Routledge.

Termeer, M. (2007). Natur unter Kontrolle – Landschaften als Bilder dritter Ordnung. In L. Engell, B. Siegert, & J. Vogl (Eds.), *Stadt – Land – Fluss. Medienlandschaften* (Archiv für Mediengeschichte, Vol. 7, pp. 171–180). Weimar: Universitätsverlag.

Tesdorpf, J. C. (1984). *Landschaftsverbrauch. Begriffsbestimmung, Ursachenanalyse und Vorschläge zur Eindämmung. Dargestellt an Beispielen Baden-Württembergs.* Berlin: Tesdorpf.

Tetens, H. (1999). Wissenschaft. In H. J. Sandkühler (Ed.), *Enzyklopädie Philosophie* (pp. 1763–1773). Hamburg: Meiner.

Theokrit. (1999). *Gedichte*. Düsseldorf: Artemis & Winkler.

Tilley, C. (1997). *A phenomenology of landscape. Places, paths and monuments* (Explorations in anthropology). Oxford: Berg.

Tillmann, E. (2016). *Bundesnaturschutzgesetz und Kulturlandschaftspflege* (Beiträge zum Raumplanungsrecht, Vol. 254). Berlin: Lexxion Verlagsgesellschaft mbH.

Toulmin, S. E. (1978). *Menschliches Erkennen I: Kritik der kollektiven Vernunft*. Frankfurt a. M.: Suhrkamp.

Treibel, A. (2004). *Einführung in soziologische Theorien der Gegenwart* (6., überarbeitete und aktualisierte Aufl.). Wiesbaden: VS Verlag.

Trepl, L. (1996). Die Landschaft und die Wissenschaft. In W. Konold (Ed.), *Naturlandschaft – Kulturlandschaft. Die Veränderung der Landschaften nach der Nutzbarmachung durch den Menschen* (pp. 13–26). Landsberg: Ecomed.

Trepl, L. (2009). Landschaftsarchitektur als angewandte Komplexitätswissenschaft? In U. Eisel & S. Körner (Eds.), *Befreite Landschaft. Moderne Landschaftsarchitektur ohne arkadischen Ballast?* (Beiträge zur Kulturgeschichte der Natur, Vol. 18, pp. 287–332). Freising: Technische Universität München.

Trepl, L. (2012). *Die Idee der Landschaft. Eine Kulturgeschichte von der Aufklärung bis zur Ökologiebewegung*. Bielefeld: transcript.

Tress, B., & Tress, G. (2001). Begriff, Theorie und System der Landschaft. Ein transdisziplinärer Ansatz zur Landschaftsforschung. *Naturschutz und Landschaftsplanung, 33*(2/3), 52–58.

Tress, B., Tress, G., & Fry, G. (2003). Potential and limitations of interdisciplinary and transdisciplinary landscape studies. In B. Tress, G. Tress, van der A. Valk, & G. Fry (Eds.), *Interdisciplinary and transdisciplinary landscape studies: Potential and limitations* (Delta series, Vol. 2, pp. 182–192). Wageningen: Delta Programm.

Tuan, Y.-F. (1979). *Landscapes of fear*. Minneapolis: University of Minnesota Press.

Türer-Baskaya, F. A. (2013). Landscape Concepts in Turkey. In D. Bruns & O. Kühne (Eds.), *Landschaften: Theorie, Praxis und internationale Bezüge* (pp. 101–113). Schwerin: Oceano Verlag.

Ueda, H. (2009). *A study on residential landscape perception through landscape image. Four case studies in German and Japanese rural communities*. Kassel: Universität Kassel.

Ueda, H. (2013). The Concept of Landscape in Japan. In D. Bruns & O. Kühne (Eds.), *Landschaften: Theorie, Praxis und internationale Bezüge* (pp. 115–130). Schwerin: Oceano Verlag.

Uekötter, F. (2007). *Umweltgeschichte im 19. und 20. Jahrhundert* (Enzyklopädie deutscher Geschichte, Vol. 81). München: Oldenbourg.

Urmersbach, V. (2009). *Im Wald, da sind die Räuber. Eine Kulturgeschichte des Waldes* (Kleine Kulturgeschichten). Berlin: Vergangenheitsverlag.

van Wezemael, J., & Loepfe, M. (2009). Veränderte Prozesse der Entscheidungsfindung in der Raumentwicklung. *Geographica Helvetica, 64*(2), 106–118.

Varela, F. J., Maturana, H. R., & Uribe, R. (1974). Autopoiesis: The organization of living systems, its characterization and a model. *Biosystems, 5*(4), 187–196. https://doi.org/10.1016/0303-2647(74)90031-8.

Vicenzotti, V. (2011). *Der »Zwischenstadt« -Diskurs. Eine Analyse zwischen Wildnis, Kulturlandschaft und Stadt*. Bielefeld: transcript.

Vicenzotti, V. (2012). Gestalterische Zugänge zum suburbanen Raum – Eine Typisierung. In W. Schenk, M. Kühn, M. Leibenath, & S. Tzschaschel (Eds.), *Suburbane Räume als Kulturlandschaften* (Forschungs- und Sitzungsberichte, Vol. 236, pp. 252–275). Hannover: Selbstverlag.

Vicenzotti, V. (2019). Die Landschaft der Zwischenstadt. In O. Kühne, F. Weber, K. Berr, & C. Jenal (Eds.), *Handbuch Landschaft* (pp. 743–753). Wiesbaden: Springer VS.

Vietta, S. (1995). *Die vollendete Speculation führt zur Natur zurück. Natur und Ästhetik*. Leipzig: Reclam.

Vilsmaier, U., & Lang, D. J. (2014). Transdisziplinäre Forschung. In H. Heinrichs & G. Michelsen (Eds.), *Nachhaltigkeitswissenschaften* (pp. 87–113). Berlin: Springer Spektrum.

Völker, H. (2004). Von der Interdisziplinarität zur Transdisziplinarität? In F. Brand, F. Schaller, & H. Völker (Eds.), *Transdisziplinarität. Bestandsaufnahme und Perspektiven. Beiträge zur THESIS-Arbeitstagung im Oktober 2003 in Göttingen* (pp. 9–28). Göttingen: Universitätsverlag Göttingen.

Waag, P. (artec – Forschungszentrum Nachhaltigkeit, Ed.). (2012). Inter- und transdisziplinäre (Nachhaltigkeits-)Forschung in Wissenschaft und Gesellschaft. artec-paper: 181. https://www. uni-bremen.de/fileadmin/user_upload/sites/artec/Publikationen/artec_Paper/181_paper.pdf. Accessed 4 Dec 2018.

Wachholz, M. (2005). *Entgrenzung der Geschichte. Eine Untersuchung zum historischen Denken der amerikanischen Postmoderne*. Heidelberg: Universitätsverlag Winter.

Wagner, P. L., & Mikesell, M. W. (Eds.). (1962). *Readings in cultural geography*. Chicago: University of Chicago Press.

Waldenfels, B. (2005). *In den Netzen der Lebenswelt* (Suhrkamp-Taschenbuch Wissenschaft, Vol. 545, 3. Aufl.). Frankfurt a. M.: Suhrkamp.

Walter, F., Marg, S., Geiges, L., & Butzlaff, F. (Eds.). (2013). *Die neue Macht der Bürger. Was motiviert die Protestbewegungen? BP-Gesellschaftsstudie*. Reinbek bei Hamburg: Rowohlt.

Wardenga, U. (1989). Wieder einmal: „Geographie heute?". Zur disziplinhistorischen Charakteristik einiger Verlaufsmomente in der Geographiegeschichte. In P. Sedlacek (Ed.), *Programm und Praxis qualitativer Sozialgeographie* (Wahrnehmungsgeographische Studien, Vol. 6, pp. 21–27). Oldenburg: BIS-Verlag.

Wardenga, U. (2001). Theorie und Praxis der länderkundlichen Forschung und Darstellung in Deutschland. In F.-D. Grimm & U. Wardenga (Eds.), *Zur Entwicklung des länderkundlichen Ansatzes* (Beiträge zur Regionalen Geographie, Vol. 53, pp. 9–35). Leipzig: Selbstverlag.

Wardenga, U. (2006). Raum- und Kulturbegriffe in der Geographie. In M. Dickel & D. Kanwischer (Eds.), *TatOrte. Neue Raumkonzepte didaktisch inszeniert* (Praxis Neue Kulturgeographie, Vol. 3, pp. 21–47). Berlin: LIT.

Warnke, M. (1992). *Politische Landschaft. Zur Kunstgeschichte der Natur*. München: Hanser.

Warren, S. (1994). Disneyfication of the metropolis: Popular resistance in Seattle. *Journal of Urban Affairs, 16*(2), 89–107. https://doi.org/10.1111/j.1467-9906.1994.tb00319.x.

Weber, M. (2011 [1919]). *Wissenschaft als Beruf* (11. Auflage). Berlin: Duncker & Humblot.

Weber, F. (2015). Diskurs – Macht – Landschaft. Potenziale der Diskurs- und Hegemonietheorie von Ernesto Laclau und Chantal Mouffe für die Landschaftsforschung. In S. Kost & A. Schönwald (Eds.), *Landschaftswandel – Wandel von Machtstrukturen* (pp. 97–112). Wiesbaden: Springer VS.

Weber, F. (2017). Widerstände im Zuge des Stromnetzausbaus – Eine diskurstheoretische Analyse der Argumentationsmuster von Bürgerinitiativen in Anschluss an Laclau und Mouffe. *Berichte. Geographie und Landeskunde, 91*(2), 139–154.

Weber, F. (2018a). *Konflikte um die Energiewende. Vom Diskurs zur Praxis*. Wiesbaden: Springer VS.

Weber, F. (2018b). Von der Theorie zur Praxis – Konflikte denken mit Chantal Mouffe. In O. Kühne & F. Weber (Eds.), *Bausteine der Energiewende* (pp. 187–206). Wiesbaden: Springer VS.

Weber, F. (2019a). Diskurstheoretische Landschaftsforschung. In O. Kühne, F. Weber, K. Berr, & C. Jenal (Eds.), *Handbuch Landschaft* (pp. 105–117). Wiesbaden: Springer VS.

Weber, F. (2019b). Stromnetzausbau und Landschaft. In O. Kühne, F. Weber, K. Berr, & C. Jenal (Eds.), *Handbuch Landschaft* (pp. 871–883). Wiesbaden: Springer VS.

Weber, F., & Jenal, C. (2018). Gegen den Wind – Konfliktlinien beim Ausbau erneuerbarer Energien in Großschutzgebieten am Beispiel der Windenergie in den Naturparken Soonwald-Nahe und Rhein-Westerwald. In F. Weber, F. Weber, & C. Jenal (Eds.), *Wohin des Weges? Regionalentwicklung in Großschutzgebieten* (Arbeitsberichte der ARL, Vol. 21, pp. 217–249). Hannover: Selbstverlag.

Weber, F., & Kühne, O. (2016). Räume unter Strom Eine diskurstheoretische Analyse zu Aushandlungsprozessen im Zuge des Stromnetzausbaus. *Raumforschung und Raumordnung, 74*(4), 323–338. https://doi.org/10.1007/s13147-016-0417-4.

Weber, F., & Kühne, O. (2019a). Die Gewinnung mineralischer Rohstoffe und Landschaft. In O. Kühne, F. Weber, K. Berr, & C. Jenal (Eds.), *Handbuch Landschaft* (pp. 843–857). Wiesbaden: Springer VS.

Weber, F., & Kühne, O. (2019b). Essentialistische Landschafts- und positivistische Raumforschung. In O. Kühne, F. Weber, K. Berr, & C. Jenal (Eds.), *Handbuch Landschaft* (pp. 57–68). Wiesbaden: Springer VS.

Weber, F., Kühne, O., Jenal, C., Sanio, T., Langer, K., & Igel, M. (2016). Analyse des öffentlichen Diskurses zu gesundheitlichen Auswirkungen von Hochspannungsleitungen–Handlungsempfehlungen für die strahlenschutzbezogene Kommunikation beim Stromnetzausbau. Ressortforschungsbericht. https://doris.bfs.de/jspui/bitstream/urn:nbn:de:0221-2016050414038/3/BfS_2016_3614S80008.pdf. Accessed 17 Oct 2018.

Weber, F., Jenal, C., & Kühne, O. (2016). Der Stromnetzausbau als konflikträchtiges Terrain. The German power grid extension as a terrain of conflict. *UMID – Umwelt und Mensch-Informationsdienst, 1,* 50–56. http://www.umweltbundesamt.de/sites/default/files/medien/378/publikationen/umid_01_2016_internet.pdf. Accessed 30 Aug 2017.

Weber, F., Roßmeier, A., Jenal, C., & Kühne, O. (2017). Landschaftswandel als Konflikt. Ein Vergleich von Argumentationsmustern beim Windkraft- und beim Stromnetzausbau aus diskurstheoretischer Perspektive. In O. Kühne, H. Megerle, & F. Weber (Eds.), *Landschaftsästhetik und Landschaftswandel* (pp. 215–244). Wiesbaden: Springer VS.

Weber, F., Kühne, O., Jenal, C., Aschenbrand, E., & Artuković, A. (2018). *Sand im Getriebe. Aushandlungsprozesse um die Gewinnung mineralischer Rohstoffe aus konflikttheoretischer Perspektive nach Ralf Dahrendorf.* Wiesbaden: Springer VS.

Weber, F., Kühne, O., & Jenal, C. (2019). Heimat und Landschaft – Zu einem eng relationierten Verhältnis. In O. Kühne, F. Weber, K. Berr, & C. Jenal (Eds.), *Handbuch Landschaft* (pp. 335–349). Wiesbaden: Springer VS.

Weber, F., Kühne, O., & Hülz, M. (2019). Zur Aktualität von ‚Heimat' als polvalentem Konstrukt – Eine Einführung. In M. Hülz, O. Kühne, & F. Weber (Eds.), *Heimat. Ein vielfältiges Konstrukt* (pp. 3–23). Wiesbaden: Springer VS.

Weith, T., & Danielzyk, R. (2016). Transdisziplinäre Forschung – Mehrwert für die Raumwissenschaften. Fünf Thesen zur Diskussion. *Nachrichten der ARL, 2,* 8–12.

Werlen, B., & Weingarten, M. (2005). Tun, Handeln, Strukturieren – Gesellschaft, Struktur, Raum. In M. Weingarten (Ed.), *Strukturierung von Raum und Landschaft. Konzepte in Ökologie und der Theorie gesellschaftlicher Naturverhältnisse* (pp. 177–221). Münster: Westfälisches Dampfboot.

Williams, R. (1973). *The country and the city.* New York: Oxford University Press.

Wimmer, C. A. (2004). Zur schönen Aussicht Typologie und Genese einer ästhetischen Erfahrung. In G. Horn (Ed.), *Wege zum Garten. Gewidmet Michael Seiler zum 65. Geburtstag* (pp. 30–36). Leipzig: Koehler & Amelang.

Winchester, H. P. M., Kong, L., & Dunn, K. (2003). *Landscapes. Ways of imagining the world.* London: Routledge.

Winiwarter, V., & Knoll, M. (2007). *Umweltgeschichte. Eine Einführung* (UTB Geschichte, Naturwissenschaften, Vol. 2521). Köln: Böhlau.

Wöbse, H.-H. (1999). Kulturlandschaft" und „historische Kulturlandschaft. *Informationen zur Raumentwicklung, 5,* 269–278.

Wöbse, H.-H. (2002). *Landschaftsästhetik. Über das Wesen, die Bedeutung und den Umgang mit landschaftlicher Schönheit.* Stuttgart: Ulmer.

Wojtkiewicz, W. (2015). *Sinn – Bild – Landschaft. Landschaftsverständnisse in der Landschaftsplanung: Eine Untersuchung von Idealvorstellungen und Bedeutungszuweisungen.* Berlin: Technische Universität Berlin.

Wojtkiewicz, W., & Heiland, S. (2012). Landschaftsverständnisse in der Landschaftsplanung Eine semantische Analyse der Verwendung des Wortes „Landschaft" in kommunalen Landschaftsplänen. *Raumforschung und Raumordnung, 70*(2), 133–145. https://doi.org/10.1007/s13147-011-0138-7.

Wylie, J. (2005). A single day's walking: Narrating self and landscape on the South West Coast Path. *Transactions of the Institute of British Geographers, 30*(2), 234–247. https://doi.org/10.1111/j.1475-5661.2005.00163.x.

Wylie, J. (2007). *Landscape.* Abingdon: Routledge.

Wylie, J. (2015). Poststructuralist Approaches: Deconstruction and Discourse Analysis. In S. C. Aitken & G. Valentine (Eds.), *Approaches to Human Geography Philosophies, Theories, People and Practices* (2nd ed., pp. 373–384). Los Angeles: Sage.

Zapatka, C. (1995). *The American landscape.* New York: Princeton Architectural Press.

Zhang, K., Zhao, J., & Bruns, D. (2013). Landschaftsbegriffe in China. In D. Bruns & O. Kühne (Eds.), *Landschaften: Theorie, Praxis und internationale Bezüge* (pp. 133–150). Schwerin: Oceano Verlag.

Zink, M. (2006). Von den Elementen zur Landschaft. In K.-H. Spieß (Ed.), *Landschaften im Mittelalter* (pp. 199–206). Stuttgart: Franz Steiner.

Zöller, R. (2015). *Was ist eigentlich Heimat? Annäherung an ein Gefühl.* Berlin: Ch. Links Verlag.

Zürn, M. (2008). Governance in einer sich wandelnden Welt – Eine Zwischenbilanz. In G. F. Schuppert & M. Zürn (Eds.), *Governance in einer sich wandelnden Welt* (Politische Vierteljahresschrift Sonderheft, Vol. 41, pp. 553–580). Wiesbaden: VS Verlag.

Zutz, A. (2015). Von der Ohnmacht über die Macht zur demokratischen Neuaushandlung. Die geschichtliche Herausbildung der Position des Planers zur Gewährleistung ‚Landschaftlicher Daseinsvorsorge'. In S. Kost & A. Schönwald (Eds.), *Landschaftswandel – Wandel von Machtstrukturen* (pp. 65–94). Wiesbaden: Springer VS.